ARROYO CENTER

T0302832

A Stated Preference Analysis of the Determinants of Unit and Soldier Operational Effectiveness

Craig A. Bond, M. Wade Markel, Olesya Tkacheva,

Richard E. Darilek, Robert G. Fix, Bryan W. Hallmark,

Henry A. Leonard, Duncan Long, Todd Nichols

Prepared for the United States Army

For more information on this publication, visit www.rand.org/t/RR1745

Library of Congress Cataloging-in-Publication Data is available for this publication.
ISBN: 978-0-8330-9701-9

Published by the RAND Corporation, Santa Monica, Calif.

© Copyright 2019 RAND Corporation

RAND® is a registered trademark.

www.rand.org

Preface

This report documents findings from the project "Assessment of Operational Performance of Army Units and Individuals." This project's purpose was to provide an assessment of the operational performance of U.S. Army units and individuals in Iraq and Afghanistan after more than ten years of counterinsurgency operations. This assessment will help the Army design its force mix with contributions from each of its components—Regular Army, Army National Guard (ARNG), and U.S. Army Reserve (USAR)—to best fulfill operational demands of the joint force commanders

The report explains the project's research method, describes the analytical results, and derives findings from that analysis. The report also provides recommendations that follow from those findings. The research described in this report should be of interest to defense and Army strategists, force planners, and resource managers.

The Project Unique Identification Code (PUIC) for the project that produced this document is HQD126411.

This research was conducted within RAND Arroyo Center's Strategy, Doctrine, and Resources Program. RAND Arroyo Center, part of the RAND Corporation, is a federally funded research and development center sponsored by the United States Army.

RAND operates under a "Federal-Wide Assurance" (FWA00003425) and complies with the Code of Federal Regulations for the Protection of Human Subjects Under United States Law (45 CFR 46), also known as "the Common Rule," as well as with the implementation guidance set forth in DoD Instruction 3216.02. As applicable, this compliance includes reviews and approvals by RAND's Institutional Review Board (the Human Subjects Protection Committee) and by the U.S. Army. The views of sources utilized in this study are solely their own and do not represent the official policy or position of DoD or the U.S. Government.

Contents

Figure and Tables

Figure

Tables

Summary

As the Army considers end strength and force structure changes in response to changes in the strategic environment, defense strategy and priorities, and fiscal pressures, understanding the attributes and competencies of units and individuals from the Army's three components becomes particularly important, especially with respect to the question of the Army's force mix. More than a decade of operational experience conducting counterinsurgency and related operations should provide ample empirical grounds for assessing relative capabilities of similar units and individuals from all three Army components. Unfortunately, no such assessment appears to have been conducted to date, nor are the data needed to permit such a study available.

Research Approach

A wealth of experience does not necessarily translate into ample, accurate, and consistent data. More than a decade of counterinsurgency experience presents almost as many challenges as opportunities for analysts. On the one hand, it provides today's veteran soldiers with ample experience on which to base an assessment. On the other hand, local differences varied considerably from year to year with respect to terrain, population, friendly forces, enemy forces, local and national (in host countries) political developments, and many other factors. Such variation makes a strictly empirical comparison, over a multiyear period, difficult. Joint doctrine refers to counterinsurgency operations as a mosaic war because of these variations. Such data are not available in any systematic form. In sum, although U.S. forces' experience and the accumulated metrics may be vast, they are insufficient to assess units' relative performance and effectiveness.

Army officers' professional experience and judgment, however, have been honed in more than a decade of combat experience. To harness this rich experience, and because usable data were not available to compare relative effectiveness, we employed a "stated preference" approach. This survey-based method was used to uncover the preferences of officers, based on their military experience and other factors, toward various attributes that described units or individuals across multiple hypothetical (yet realistic) scenarios.

Choice Experiments

We used a "stated preference" methodology known as a *choice experiment*. Based in microeconomic consumer theory, the general idea is to present respondents with one or more choices over a set of discrete options (here, either units or individuals) that vary across key attributes and ask the respondent to choose their preferred option. Here, the context is to select a notional unit or individual to perform a mission or function specific to that scenario. Respondents are then presented with a series of choices in which they are to select the option that is most appropriate to the stated mission or function within the overall context of an ongoing campaign. Potential units and individuals vary across important characteristics, including component, unit type, previous operational experience, and availability, among others. The research design assumed respondents would rely on their operational experience in assessing the relative importance of these characteristics and made their choices to maximize success in the notional campaign described in the scenario.

Using appropriate experimental designs for the survey and statistical techniques for the analysis, the researcher is then able to use the responses to estimate an equation that gives insight into the preferences of the sample. In particular, this equation can be used to infer (1) the relative importance of each attribute to respondents relative to a baseline and (2) the willingness of respondents to trade off one attribute for another. Because we assume that respondents would prefer the most effective unit subject to the constraints implied by each choice occasion, the estimated equations provide evidence about the *perceived* operational effectiveness of units or individuals across different scenarios from knowledgeable experts (i.e., experienced military officers). The survey is experimentally designed such that respondents must trade off the values of attributes against one another to make a choice.

For example, suppose the scenario involved choosing a unit for a counterinsurgency mission, and the distinguishing characteristics between units at one choice occasion were training and type of experience. Assume each of these characteristics had two potential levels (for instance, three months and six months for training, and security and counterinsurgency for type of experience). Because both training and counterinsurgency experience are likely to enhance effectiveness for this mission, we would expect that an informed respondent would prefer a unit with six months training and counterinsurgency experience if given the choice between all four possible unit types (i.e., units with [1] three months training and security experience, [2] six months training and security experience, [3] three months training and counterinsurgency experience, and [4] six months training and counterinsurgency experience). However, offering such a choice alone does not allow for inference about the relative strength of the preferences for, say, counterinsurgency experience relative to three extra months of training.

To overcome this problem, an experimental design might offer a choice only between unit A, which has three months of training and counterinsurgency experience, and unit B, which has six months of training and security experience. If the respondent selects unit A, we would infer that counterinsurgency experience (relative to security experience) is more valuable to that individual than three extra months of training. This inference can be made because the respondent had the chance to choose the extra three months of training but did not. In this case, the experimental design purposefully eliminates the globally preferred unit to provide statistical information about relative values of the characteristics. In other words, the experimental design functions as a constraint set that forces respondents to make trade-offs between relevant attributes. In a world of scarce resources, such trade-offs are the norm.

Value of Attributes

Given enough responses from enough individuals and an appropriate experimental design, it is possible to statistically infer the relative value attached to relevant unit and individual characteristics by the sample participants by estimating an equation that predicts the observed (stated) choices as a function of the levels of characteristics presented to each respondent. The results reflect perception and judgment of the respondents; they are not necessarily verifiable through objective, quantitative means. In addition, hypothetical and other biases may affect the results. However, proper survey design and focus grouping can minimize these problems.

Furthermore, in the context of unit and individual operational effectiveness for which other approaches to establishing value are impossible or prohibitively difficult (e.g., randomized control trials deployed during the course of a war), such an approach provides a means of establishing relative value. Relative value is indeed the important issue. In this case, one aspect of the analysis is a foregone conclusion. Past discourse on the issue indicates that Regular Army officers, based on their operational experience and other factors, will typically prefer Regular Army units and individuals for the more complex missions, a preference not necessarily shared by their Army National Guard (ARNG) counterparts. Our results are generally congruent with this perception (which is, in itself, and important validity check). The interesting question, however, is *how much* Regular Army respondents are willing to give up in terms of other important characteristics to get the capability that they want.

Survey Sample

We elicited data for the stated preference analysis through a web-based survey of serving Regular Army, ARNG, and U.S. Army Reserve (USAR) colonels and general officers with a background in the following branches: infantry, armor, field artillery, special forces, engineers, military police (MP), and logistics. Of the approximately 2,400 individuals who were invited to participate, only 217 completed the survey. Table S.1 breaks the sample down by the most responsible position respondents held when deployed and the average number of months for which they were deployed. As indicated in Table S.1, there were significant differences in Regular Army, ARNG, and USAR respondents' average level of experience. The former averaged 34.6 months deployed,

Table S.1
Sample Distribution by Most Responsible Position Held When Deployed and Average Months Deployed

	Active		Guard		Reserve	
	Count	Average Months Deployed	Count	Average Months Deployed	Count	Average Months Deployed
Commander/deputy commander at division or higher echelon	7	37.4	6	16.0	2	22.5
Chief of staff at division or higher echelon	4	44.3	3	23.0	1	12.0
Operations officer at division or higher echelon	3	33.3	3	10.7	2	33.0
Brigade combat team commander/deputy commander	7	30.3	11	23.6		
Special forces group commander			1	40.0		
Other brigade-level command	7	38.3	7	20.1	6	20.5
Maneuver/fires battalion commander	13	34.8	10	24.1		
Other battalion-level command	12	31.7	7	14.3	2	23.0
Transition team commander	1	40.0	9	21.3	5	19.4
Force management staff officer					2	8.0
Staff officer at division or higher echelon	13	34.5	13	19.0	16	30.2
Brigade/battalion-level staff officer	4	25.5	12	16.6	7	13.0
Other	2	58.0	12	15.1	7	13.9
Total	**73**	**34.6**	**94**	**19.1**	**50**	**21.1**

while the latter respondents had been deployed for an average of 19.8 months. There were probably significant differences in the nature of their operational experiences as well as the extent thereof. For example, Regular Army combat brigades deployed to Iraq and Afghanistan predominantly conducted either close combat or counterinsurgency operations. In contrast, other RAND research has indicated that only about nine of the 45 ARNG maneuver brigades deployed to Iraq and Afghanistan between 2003 and 2010 performed in a counterinsurgency role.

To make credible inferences about the population of Army colonels and general officers, the sample used should be random and thus representative of that population. Given the constraints we faced in the administration of the survey (especially an inability to issue follow-up requests for participation), we cannot be sure that our sample is truly random and representative; that is, there may some bias in the results if respondents hold substantially different preferences (and thus patterns of survey response) than nonrespondents. This self-selection bias is possible with all survey work and cannot be empirically tested without additional information on the nonrespondent population. While we have no reason to believe that systemic bias is likely in this application, readers are cautioned that extrapolation to all Army colonels and officers may not be appropriate if preferences of the nonrespondents are substantially different than those in our sample.

Findings

Findings fall into one of three categories: unit capabilities, individual capabilities, or methodology. Our survey results suggest that the Regular Army respondents believe that the Army incurs additional risk through employing ARNG maneuver units in lieu of Regular Army units, especially for higher-risk missions. When employment of a Regular Army maneuver unit relative to an ARNG alternative would involve the disruption of higher-level operational plans or sacrificing continuity in the area of operations under low- and moderate-threat scenarios, Regular Army officers tend to find this risk acceptable (i.e., they would choose the ARNG unit). This result does not apply to higher-threat environments, however. In this case, Regular Army respondents were willing to adjust the overall campaign plan to ensure they could employ a Regular Army unit.

Respondents seem to find fewer risks in employing ARNG and USAR enablers—units that support maneuver units in the conduct of combat operations—at the battalion and company level. With respect to individual staff officers—we did not explore preferences with regard to supervisory positions or positions in such ad hoc units as advisory and transition teams—component mattered only in positions concerned with the day-to-day monitoring and management of military operations. Even there, however, Regular Army officers' preference for Regular Army candidates was trumped by the value of relevant civilian-acquired skills for ARNG and USAR soldiers.

Unit Capabilities

Respondents were to pick the best unit to perform a certain mission in a given area of operations in the overall context of a counterinsurgency campaign. Units varied with respect to a number of characteristics, including but not limited to unit type, component, experience in country, manning levels with respect to key systems, and so forth. Scenarios varied mostly with respect to the threat level.

We conducted five types of unit surveys:

- *Brigade combat teams (BCTs)* are the Army's primary maneuver forces. The term *maneuver* indicates that they are responsible for conducting combined arms operations in a given area of operations. During the wars in Iraq and Afghanistan, they comprised about 4,000 soldiers, organized into several different types of battalions (e.g., reconnaissance, maneuver, fires, and sustainment). BCTs only reside in the Regular Army and the ARNG.
- *Maneuver battalions* are, like a BCT, responsible for conducting combined arms operations within a given—albeit smaller—area of operations. Maneuver battalions generally comprise companies of a single unit type (e.g., tanks and mechanized infantry), although battalion headquarters are responsible for integrating effects provided by other arms and the services. Maneuver battalions generally have around 700 soldiers.
- *Engineer battalions* are enabler units. Under the Army's modular organization, an engineer battalion—not to be confused with a *brigade engineer battalion*—is not a single unit containing several companies, but rather a command-and-control headquarters responsible for supporting, synchronizing, and integrating the efforts of attached engineer companies. The type of engineer companies attached can differ according to the mission and operational situation, but generally include horizontal construction, vertical construction, and route clearance in a counterinsurgency scenario. Engineer battalions as described are found at echelons above brigade.
- *Engineer companies.* We limited this analysis to route clearance companies. Route clearance companies are responsible for locating and neutralizing improvised explosive devices and other obstacles emplaced along routes used by other U.S. and multinational forces and civilian traffic. They generally comprise fewer than 200 soldiers.
- *MP companies* provide law enforcement support, in addition to other capabilities. For the purposes of this analysis, however, we focused on another key MP mission: route security. Like engineer battalions and companies, MP companies are an enabler formation.

Army Officers from the Regular Army and ARNG Differed Starkly in Their Perceptions of the Relative Effectiveness of Regular Army and ARNG Maneuver Units

Regular Army respondents accorded greater weight to units' Regular Army status than to almost any other factor—except for operational risk—especially in determining which maneuver battalions or BCTs to employ. For maneuver battalions, this factor weighed more heavily than the nature or degree of units' prior experience in theater, units' amount of predeployment training time, or even the unit type (combined arms or infantry) in a high-threat environment. We did not conduct a similar analysis at the company or platoon level. In contrast, ARNG respondents accorded relatively little weight to component status, although, as previously indicated, their operational experiences typically differed significantly from their Regular Army counterparts. Army respondent preferences reflect underlying beliefs, based on operational experience and other factors, about the relative effectiveness of Regular Army and ARNG maneuver forces. Those beliefs differ substantially, as did their level of recent operational experience.

These differing perspectives inform Army decisions about force structure and resources. Those decisions can lead to internecine bureaucratic conflict, as the recent controversy over the Army's proposed Aviation Restructuring Initiative (ARI) demonstrates.[1] In fact, these differing perspectives and the resulting tension have been an issue since the founding of the country. As established in the U.S. Constitution, the National Guard (Army and Air) report to their governors in peacetime, but can be "called forth" (mobilized or federalized) by the President to conduct federal missions. Only when so federalized do National Guard units fall completely under the administrative and operational control of the U.S. Departments of Defense, Army, and Air Force. This dual chain of command has been a historic source of confusion and friction but is not the focus of this study.

---

[1] The ARI was a 2013 plan to rapidly reduce Army aviation costs and spending in response to the 2011 Budget Control Act. ARI included the elimination of nearly 700 aircraft from the Regular Army and the consolidation of all Apache attack helicopters in the Regular Army, where they would be dual-missioned for both the attack and armed reconnaissance roles. Under ARI, the ARNG and USAR would transfer their Apaches to the Regular Army in exchange for Black Hawks. Advocates for the plan, including the Office of the Secretary of Defense, cited Regular Army formations' superior responsiveness and cost-effectiveness. The ARNG and its advocates sought to retain Apaches in fulfillment of their stated role as the Army's principal combat reserve. The controversy over this decision led to the establishment of the National Commission on the Future of the Army. See U.S. House of Representatives, *Howard P. "Buck" McKeon National Defense Authorization Act for Fiscal Year 2015: Report of the Committee on Armed Services House of Representatives on H.R. 4435 Together with Additional Views*, Washington, D.C.: Committee on Armed Services, H.R. 113-446, May 13, 2014, pp. 199–200; see also National Commission on the Future of the Army, *Report to the President and the Congress of the United States*, Washington, D.C., January 28, 2016, p. i.

Operational Risk Was the Most Important Factor in Selecting BCTs

Minimizing disruptions to the overall campaign plan weighed more heavily than any other factor for all Army respondents. Regular Army officers chose Regular Army BCTs, but not if doing so would disrupt the overall campaign plan. Regular Army were willing to accept ARNG BCTs in low- and moderate-threat scenarios, if selecting an alternative would force the overall commander to defer a planned counteroffensive. In the high-threat scenario, however, Regular Army respondents were willing to disrupt or adjust the overall campaign plan to employ a Regular Army BCT.

Continuity Was Generally the Most Important Factor in Selecting Maneuver Battalions

Similarly, maximizing continuity proved to be the most important factor in respondents' choice of maneuver battalions. *Continuity* was defined as the potential to remain in an assigned area of operations over an extended period of time. Respondents from both components preferred units that could remain longer to those with less availability. Even Regular Army respondents preferred maneuver battalions with a high degree of continuity to any other factor—including training, component status, and prior experience in theater—in low- and moderate-risk scenarios. In high-threat environments, however, it was more important to Regular Army officers to have a Regular Army unit than it was to maximize continuity.

In Combination, Other Factors Could Outweigh Preferences for Regular Army Maneuver Units

All things being equal, Regular Army respondents preferred to employ Regular Army maneuver units, especially for higher-risk missions. All things are seldom equal, however, and decisionmakers must balance a number of competing factors in assigning units to particular operational environments. In these choice experiments, various conditions could combine to cause Regular Army respondents to prefer ARNG maneuver units to Regular Army units. Considerations of operational risk and continuity could trump Regular Army respondents' preference for Regular Army maneuver units in low- and moderate-threat scenarios. In combination with other factors, it could even outweigh component status in high-threat situations. For example, in the maneuver battalion survey, results indicate that Regular Army respondents would have preferred ARNG combined arms battalions with six months of predeployment training and the ability to remain in the area of operations for six months or more to Regular Army infantry battalions that might have to depart the area of operations later without being replaced in the high-threat scenario.

Respondents Considered Key Leaders' Experience and Skills Critical to Units' Operational Effectiveness

During the interviews and focus groups in support of developing the survey instruments, respondents identified key leaders' proficiency as a critical determinant of units' operational effectiveness. In this sense, leadership referred to key leaders' experience and skills synchronizing and integrating combat operations, not their inherent personal qualities such as integrity or charisma. Respondents were generally left to factor their assumptions about the level of key leaders' experience and skills in the Army's three components into their weighting of the relative value of component status. While we did not include leadership as a variable in most of our scenarios, more than 40 percent of the free responses submitted highlighted the importance of leadership in establishing units' operational effectiveness.

Regular Army Respondents' Preferences for Regular Army Enablers Was Less Pronounced

While Regular Army respondents demonstrated a clear, consistent, and unambiguous preference for Regular Army maneuver units, especially in high-risk situations, their preference for Regular Army engineer units and MP companies was at least somewhat ambiguous. This result may be a function of the relatively small sample of engineers (11) who took the unit survey or the echelons being considered (battalion and company). It should also be noted that the enabler units in question were largely providing a single operational capability in support of maneuver units, in contrast to the maneuver units that had responsibility for integrating the full range of military capabilities into operations in an area of operations. For whatever reason, Regular Army respondents were more willing to contemplate employing ARNG and USAR enablers than they were to consider using ARNG maneuver units.

Individual Capabilities

The individual survey asked respondents to select staff officers to fill positions in a 500-billet combined headquarters overseeing military operations in a medium-sized counterinsurgency operation. Respondents were to select candidates to serve as a desk officer in the operations directorate (C-3), the contract manager in the programs and resources directorate (C-8), and a counternarcotics planner in the directorate of strategic plans and policy (C-5). Candidates varied with respect to component, operational experience, branch, and civilian-acquired skills.

Regular Army Respondents Preferred to Employ Regular Army Officers in Operations Positions

In general, respondents were indifferent to component in selecting individual staff officers. The one exception to this rule was the selection of officers for positions in the operations directorate. Regular Army respondents tended to choose the Regular Army candidates for this position, which placed a premium on understanding military operations and navigating standard Army processes.

All Respondents Preferred Candidates with Relevant Civilian-Acquired Skills

All Army respondents preferred ARNG or USAR candidates with civilian skills relevant to the position in question to candidates who lacked such skills, regardless of the candidate's component. For example, an ARNG or USAR candidate who served as a prosecutor in civilian life was preferred as a counternarcotics planner to a notional Regular Army officer with little or no related experience. We note that all ARNG and USAR soldiers cannot be presumed to possess civilian skills that are relevant to military operations or available to the degree needed.

Continuity Was Important, but Not Dispositive

In the individual survey, we denominated opportunity costs in terms of continuity. Assuming that frequent rotation of personnel impaired organizational efficiency and imposed costs, we treated candidates available for short periods—six or fewer months— as having high opportunity costs and those available for longer periods—for example, a year—as having low opportunity costs. Naturally, respondents preferred individuals available for longer periods. This characteristic was less important to respondents than ARNG and USAR candidates' civilian-acquired skills, however, and less important to Regular Army respondents than component in selecting candidates for the operations directorate position.

Methodological Findings

The central question for this study lay in determining how important a unit's or individual's component was in determining operational effectiveness relative to other factors. To do so, we adapted stated preference methods—an approach developed to support nonmarket valuation and benefit cost analysis in the context of environmental regulation—to a new context. This adaptation was even more challenging because operational Army leaders seldom, if ever, are exposed to the relative monetary costs of various operational options.

This study demonstrates that it is possible to employ stated preference methods to quantify the value of characteristics in a consistent way (e.g., that characteristic A was twice as important as characteristic B, relative to a baseline characteristic). The importance of inducing trade-offs for respondents through the experimental design, thus allowing for estimation of relevant coefficients of the statistical model, is a key takeaway for future use of this method in military contexts. Ultimately, however, the salience of the resulting findings depends on the confidence that decisionmakers place in respondents' professional judgment, recognizing the hypothetical nature of the exercise.

Recommendations

The results of the stated preference exercise suggest that, despite a decade of shared combat experience, the importance of component differs between Regular Army and reserve component officers, especially for maneuver units. Regular Army officers generally perceive component as a relevant factor in choosing between units, all else equal, while Army National Guard and Army Reserve officers do not. These results, however, do not carry over to enabler units and individual positions, and the preference for Regular Army units is not absolute in that some ARNG maneuver units with particular attributes may be preferred to Regular Army maneuver units with different attributes. Although the exercise cannot explain the reasons for these differences in preferences, possible explanations include component pride and the fact that Regular Army respondents, on average, had significantly different responsibilities and experiences in their more-frequent operational deployments than did their ARNG counterparts. As a result, the "component" attribute of the exercise may be used as a proxy for unit experience, proficiency, and effectiveness.

Operationally, Regular Army officers appear to believe that employing ARNG maneuver units in lieu of Regular Army units in counterinsurgency operations incurs significant additional risk at the tactical level. Nonetheless, our analysis also suggests that additional risk remains within acceptable levels in low- to moderate-threat environments if Regular Army maneuver units are unavailable or better tasked to other missions. The risk of employing ARNG maneuver units exceeds acceptable levels in high-threat environments. Some combination of more extensive annual and predeployment training than is currently conducted would probably be necessary to reduce those perceived risks in future counterinsurgency operations. As for the employment of ARNG and USAR individuals, officers with civilian skills relevant to the positions they are to fill are considered highly desirable in a range of higher-level staff jobs in such contingency headquarters as Multi-National Force—Iraq or the International Security Assistance Force. Note that we are not talking about staff jobs at the corps, division, or brigade level. In overview, our recommendations generally follow a pattern of explicitly assessing the risk associated with different operational requirements, preserving sufficient Regular Army capacity to meet high-risk requirements and taking measures to mitigate the residual risks of employing ARNG and USAR forces in lower-risk contexts.

We did not explicitly consider the question of fiscal cost. Costs were considered in terms of operational risk instead. Effectiveness is only one part of the larger issue in developing the Army's force structure and readiness posture, however. Another is financial cost. Another RAND Arroyo Center analysis investigated the issue of relative financial costs.[2] That study determined that it can be less expensive to achieve a specified sustained level of output using Regular Army units than it would be to produce the same level of sustained capacity using ARNG or USAR units, at least for certain highly capitalized formations (e.g., armored BCTs [ABCTs] and attack helicopter battalions) under current rotational deployment policies regarding the frequency with which Army forces and individuals, by component, may be deployed. That analysis is, of course, sensitive to those rotational deployment policies and unit materiel and training costs.

With these considerations in mind, we make the following recommendations.

Consider Increased Pre-Employment Training for ARNG Forces Conducting Counterinsurgency Operations

All other things being equal, Regular Army officers preferred Regular Army units in all situations. They were, however, willing to employ ARNG maneuver forces in low- to moderate-risk contexts to mitigate operational level risk. This trade-off would incur greater tactical risk, even in these lower-risk scenarios. Regular Army officers apparently felt that increased preparation could adequately prepare ARNG maneuver units for these lower-risk scenarios. In these choice experiments, increased preparation took the form of either six months of predeployment training focused on counterinsurgency or in-country experience conducting counterinsurgency operations in a lower-risk environment. Our survey did not provide enough information to conclude that any amount of predeployment training alone could completely eliminate the risk Regular Army respondents believed was inherent in employing ARNG maneuver forces for higher-risk missions. Analysis of survey results did indicate that six months of such training—a period exceeding that prescribed by current policy—was probably insufficient in and of itself.

Our analysis indicates that, if it is necessary to employ ARNG maneuver battalions and brigades to conduct counterinsurgency operations, some combination of additional annual training days, longer periods of mobilization, and more-robust predeployment training may be necessary. Alternatively, policymakers might decide to accept the risk associated with deploying less well-trained Army units to lethal environments. Finally, the Army might also consider deploying ARNG and USAR units in platoons and companies rather than battalions, brigades, and divisions to reduce requirements for predeployment training.

---

[2]   Joshua Klimas, Richard E. Darilek, Caroline Baxter, James Dryden, Thomas F. Lippiatt, Laurie L. McDonald, J. Michael Polich, Jerry M. Sollinger, and Stephen Watts, *Assessing the Army's Active-Reserve Component Force Mix*, Santa Monica, Calif.: RAND Corporation, RR-417-1-A, 2014.

Assess the Utility of Increasing Key ARNG and USAR Leaders' Opportunities to Accrue More Operational Experience and Assign Regular Army Leaders to Command and Staff Billets in ARNG and USAR Units

Respondents from all three components identified key leader's ability to synchronize and integrate combat operations as a critical determinant of unit effectiveness. Increasing ARNG and USAR key leaders' opportunities to acquire operational experience could produce results disproportionate to the investment. Options for doing so could include encouraging midgrade Regular Army leaders' transition to the ARNG or USAR, mobilizing key upwardly mobile ARNG and USAR officers to deploy with Regular Army units while fulfilling developmental roles, and assigning Regular Army leaders to key command and staff billets in ARNG and USAR units. Some measures might require legislative approval, as well as a slightly larger Regular Army force, to fulfill the additional command and staff billets in ARNG and USAR units. Moreover, all such options require the ARNG and USAR to manage their key leaders closely to ensure that those with the greatest operational responsibilities have accumulated the most relevant experience. As with other recommendations, further analysis would be required.

Explore Options for Increasing Access to ARNG and USAR Soldiers with Relevant Civilian-Acquired Skills

Respondents preferred soldiers with civilian skills relevant to the positions for which they were being considered to those without, regardless of the soldier's component. This dynamic applied even in operations directorate positions primarily concerned with monitoring and managing military operations. The Army does not assign soldiers to military billets based on their civilian-acquired skills, or track soldiers' civilian skills, regardless of component. Previous U.S. Department of Defense efforts to catalog military personnel's civilian-acquired skills have not proven fruitful to date. The Army could increase its ability to identify potential operational requirements for soldiers with particular talents and incentivize soldiers with those skills to volunteer to fill positions in which those skills would be useful.

Conclusion

The study's purpose was to assess the relative operational effectiveness of Army units and individuals from all three Army components to most effectively fulfill operational demands in a counterinsurgency environment. Regular Army maneuver units, given their greater opportunities and resourcing for peacetime training and readiness, are probably required to meet high-threat operational requirements. Regular Army respondents also preferred to employ Regular Army units in low- and moderate-threat environments, but felt that ARNG maneuver units could be employed in such environments at acceptable levels of risk if employing Regular Army units would disrupt higher operational priorities. Similarly, Regular Army respondents generally concluded that ARNG and USAR enablers could fulfill their roles without incurring undue additional risk.

After more than a decade of deployments to two theaters of war, Regular Army officers perceive more differences in operational performance and risk between full- and part-time forces than do their less-experienced ARNG counterparts, especially for maneuver forces conducting high-risk operations. This assessment assumes that respondents completed the stated preference exercise on the basis of their professional experience and judgments about the best course of action for the scenario described. These findings and the associated recommendations rest on the judgment of a relatively small sample of highly experienced Regular Army colonels and general officers. Their ARNG counterparts perceive fewer differences, at least on the subject of the relative effectiveness of Regular Army and ARNG maneuver units. Additional analysis would be required to determine the extent to which these perceived differences should inform the allocation of the Army's capabilities among its three components, as well as how their units are operationally employed.

There is also the question of the degree to which policymakers can or should extrapolate from these findings and recommendations specific to a counterinsurgency environment to other contexts. Caution is certainly warranted in applying lessons learned in one context to a substantially different one.

Acknowledgments

We would like to thank the late MG John Rossi, BG Frank Muth, and Tim Muchmore of the Army Quadrennial Defense Review Office for sponsoring this research. Robert Simmons of the Army Research Institute helped guide us through the U.S. Army's survey approval process and provided valuable input on the survey itself. We also wish to thank COL Randall Cheeseborough and Anna Waggener of the Army War College for helping coordinate the focus groups we used to validate the survey instruments. Thanks also go to Terrance Enright of the Army G-8 for helping us obtain the necessary permissions to administer the survey via Army Knowledge Online. We also thank Jill Luoto, Gian Gentile, and Stephen Biddle for helpful comments that greatly improved the clarity and quality of the report. Finally, we could not have completed our research without help the Army fellows who helped pilot the survey, Melissa Bradley and her colleagues in the RAND Corporation's Survey Research Group, and Gina Frost, who shepherded this manuscript through the RAND Arroyo Center's publications process.

Abbreviations

ABCT	armored brigade combat team
ARI	Aviation Restructuring Initiative
ARNG	Army National Guard
BCT	brigade combat team
C2	command and control
CJTF	Combined Joint Task Force
CTC	Combat Training Center
DoD	U.S. Department of Defense
FAR	Federal Acquisition Regulations
FM	Field Manual
FOB	forward operating base
HBCT	heavy brigade combat team
HMMWV	high-mobility, multi-purpose wheeled vehicle
IBCT	infantry brigade combat team
IED	improvised explosive device
MP	military police
MRAP	mine-resistant ambush protected
MRE	Mission Rehearsal Exercise
OR	operational readiness
RSTA	reconnaissance, surveillance, and target acquisition
SFATT	Security Force Assistance Transition Team
SECFOR	security force
USACE	U.S. Army Corps of Engineers
USAR	U.S. Army Reserve

CHAPTER ONE

Introduction

Background

The U.S. Army comprises three components: the Regular Army, the Army National Guard (ARNG), and the U.S. Army Reserve (USAR). The Army combines forces from each of its components into larger organizations to accomplish assigned missions, but empirical evidence of expected relative capabilities of units and individuals from its components is largely nonexistent. As the U.S. Army adjusts its force structure in response to changes in strategic and fiscal guidance, understanding those attributes and competencies becomes particularly important, especially with respect to the question of the Army's force mix between components.

Debates on this issue generally break down between those who believe that ARNG and USAR units can provide equivalent capability at lower cost and those who believe that Regular Army forces are required to provide certain key capabilities at acceptable cost and risk.[1] Testimony from the former Chief of the National Guard Bureau, GEN Frank Grass, illustrates the argument made by the former group:

> If mobilized, these units can achieve Brigade Combat Team [BCT] level proficiency after 50–80 days of post-mobilization training. When deployed for operational missions Guard and Active Army units are indistinguishable. Army Guard Brigade Combat Teams will not replace early deploying Active Army Brigade Combat Teams in their overseas "fight tonight" missions.[2]

---

[1] Andrew Feickert and Lawrence Kapp, *Army Regular Army (AC)/Reserve Component (RC) Force Mix: Considerations and Options for Congress*, Washington, D.C.: Congressional Research Service, R43808, December 5, 2014, pp. ii.

[2] Frank Grass, "Statement by General Frank J. Grass, Chief, National Guard Bureau, Before the Senate Armed Services Committee, 2nd Sess., 113th Cong., on Army Total Force Mix," Washington, D.C., April 8, 2014. See also Frank Grass, "Authorities and Assumptions Related to Rotational Use of the National Guard," memorandum for Chief of Staff of the Army," May 31, 2013; Office of the Vice Chairman of the Joint Chiefs of Staff and Office of Assistant Secretary of Defense for Reserve Affairs, *Comprehensive Review of the Future Role of the Reserve Component*, Washington, D.C.: U.S. Department of Defense, 2011.

Exponents of this argument base their position on the Army's heavy reliance on ARNG combat forces in various roles in more than a decade of conflict in Afghanistan and Iraq. One observer went so far as to say "[a]fter 12 years of hard fighting in Afghanistan and Iraq, however, where regulars, reservists, and guards fought side by side and were mostly treated as interchangeable, it's almost impossible to make the effectiveness argument anymore."[3] Regular Army officers are less sanguine, noting that the degree to which ARNG and USAR forces can substitute for Regular Army forces depends heavily on the complexity and risk of the missions they are called on to perform. They argue that ARNG maneuver forces did not, for the most part, perform missions of equivalent complexity and risk in recent operations, relative to how Regular Army maneuver forces were employed.[4]

Assessing relative effectiveness by component can inform the Army's appropriate force mix. If ARNG and USAR units can provide the needed capability (including availability, readiness, and proficiency) at lower cost, it would suggest that the Army could assign these forces to the ARNG and USAR. If, however, the ARNG and USAR provide less capability for less cost, it would shift the balance to retaining Regular Army capacity. Unfortunately, as Andrew Feickert and Lawrence Kapp of the Congressional Research Service note, "There does not appear to be any systematic assessment of unit performance during the wars in Iraq and Afghanistan that would be suitable for comparing unit effectiveness between AC [active component] and RC [reserve component] units."[5]

Unit effectiveness is not the only relevant question. The Army relied heavily on individual ARNG and USAR soldiers to staff contingency headquarters, help train Afghan and Iraqi forces, and augment Regular Army forces over the course of these wars. Many of these billets do not exist in standard unit authorization documents, but must be filled from within the Army's inventory in the event of conflict. If the Army must retain enough Regular Army soldiers to fill these roles, over and above manning requirements for extant organizations, it either requires a larger Regular Army or reduces the number of units that can be fielded within a fixed end strength. If ARNG and USAR soldiers can fill some of these roles, it increases the Army's options.

To enable the Army to help make these assessments, the Army Quadrennial Defense Review Office asked the RAND Corporation's Arroyo Center to provide an assessment of the operational performance of Army units and individuals from all components after more than a decade of war. The purpose of this assessment was not

---

[3] Sydney J. Freedberg, Jr., "Active vs. Guard: An Avoidable Pentagon War," *Breaking Defense*, June 28, 2013. See also Phillip Carter and Nora Bensahel, "Reboot: Why the Army's Plan to Cut 80,000 Troops Doesn't Go Nearly Far Enough," *Foreign Policy*, June 26, 2013.

[4] Sydney J. Freedberg, Jr., "National Guard Commanders Rise in Revolt Against Active Army; MG Rossi Questions Guard Combat Role," *Breaking Defense*, March 11, 2014.

[5] Feickert and Kapp, 2014, p. 28.

simply to establish which component provided "better" units or individuals, but rather to assess how much difference typically existed between similar units and individuals from different components, with respect to similar missions or jobs.

Obstacles to the Research Effort

The study team confronted a number of major obstacles in developing a research design. When assessing individual or unit proficiency, analysts would prefer to compare identical units' performance of identical tasks or missions under identical circumstances. Needless to say, conditions of such uniformity rarely exist over several years in combat in two countries and spanning two administrations from different political parties. Even similar units from the same component may be assigned different missions or face significantly different conditions on the ground. To the extent that data on those conditions—to include the plans, disposition, and conditions of enemy forces—are reasonably comprehensive and the sample size is sufficiently large, it may be possible to use statistical techniques to control for such variations.

None of these conditions was met with regard to U.S. unit performance in counterinsurgency operations. The war—or wars—proceeded in different ways, against different enemies, in different places, at different times. U.S. forces actively sought—with some success—to gather information on enemy organization, intentions, and dispositions. To the extent those data are available, it is not easy to translate them into a form that supports quantitative analysis. Surprisingly, it is similarly difficult to establish information about U.S. units' day-to-day dispositions after the fact. Synthesizing that information into a data set that would allow statistical comparison of unit-level effectiveness might well yield valuable insight, but would require considerable time and effort. We sought instead to assess individual officers' preferences for units or individuals with varying characteristics across different scenario types using surveys and stated preference analysis.

Research Approach: Stated Preference Methodologies

Stated preference analysis is a group of methodologies that is used to estimate the underlying preferences of a sample or population of individuals. Based in microeconomic consumer theory, the general idea is to present respondents with one or more choices over a particular good or set of goods, and ask the respondent to choose which is preferred.[6] Using appropriate experimental designs for the survey, the researcher is

---

[6]   More specifically, the technique is grounded in utility theory, and the estimated equations are typically indirect utility functions.

then able to use the responses in aggregate to estimate an equation that gives insight into how the goods (and possibly their attributes) are valued. The term "stated preference" is used to distinguish these techniques from "revealed preferences," in which individuals or other agents actually act on their preferences in a marketplace or other setting. A primary benefit of the "stated preference" technique is the ability to present choices that may be outside of the range of observable revealed choices; however, the hypothetical nature of the choices implies that one cannot be certain that respondents would behave as implied in the real world.

The experimental design is an important part of any stated preference analysis, as it provides the structure necessary to estimate the equation that describes the underlying preferences. In particular, the survey questions are designed to induce trade-offs between goods and/or their attributes, mimicking a constrained situation. In other words, by design, respondents may be forced to make a choice between what otherwise might be their second- or sixth-best options, relative to an unconstrained situation.

A simple example can help illustrate. Consider a sample of respondents who are asked to choose between an apple and an orange. All else equal (including the potential price of each piece of fruit), assume each individual in the sample prefers apples.[7] However, to determine the relative strength of this preference, a researcher could vary the relative prices of apples and oranges across the sample and estimate the probability of choosing each. The variation in prices alters the opportunity cost of choosing an apple over an orange across respondents, and thus allows the researcher to determine the degree to which the average respondent is willing to trade off dollars (a proxy for all other goods) for apples and oranges.

A similar, although more complex, experiment was done for this report in the context of unit and individual operational readiness. Each respondent completed a survey on one type of maneuver unit, enabler unit, or the individual survey based on their qualifications and operational experience. Within each category, the sample of officers was asked to make a series of choices from among subsets of units or individuals with varying characteristics for each scenario. For unit surveys, each scenario differed according to operational environment, threat level, and incoming unit mission. For the individual survey, different scenarios represented alternative job descriptions.

After a description of these elements for each scenario, the appropriate subsample of officers was presented a series of choices between units or individuals that differed in terms of their descriptive characteristics. For example, for BCTs, the characteristics included in-country experience, Army component, unit type, postmobilization training levels (for ARNG and USAR units), and the level of operational risk that would accrue to the commander's overall plan (but not the unit's mission itself), ostensibly because of relative scarcity of resources (or the lack thereof). This latter attribute served

---

[7] This would not, in general, be known to the researcher.

a similar role to the price attribute in the apples and oranges example in that it represented the opportunity costs to the overall plan of choosing a particular unit.

For each choice occasion, respondents had to trade off the set of unit-specific attributes against the operational risk to the overall plan that would be induced by that choice. Statistical techniques were then used to estimate an equation that represented the strength of preferences toward each attribute in each scenario.

Analytic Objective

The objective of this analysis was to infer assessments made by officers with different levels of operational experience of the relative importance of the attributes described in our choice scenarios. We assumed that respondents' preferences in the choice experiments would reflect these assessments. Implicit in our assumptions is that respondents to our stated preference surveys will choose the unit or individual that he or she believes is best suited to perform the mission described in the scenario given the alternatives available, as described in terms of their relevant characteristics. One of the more important characteristics was the opportunity cost of making any choice. In so doing, we provide evidence as to the knowledgeable experts'—military officers' with different levels of operational experience—perceptions of the relative operational effectiveness of units or individuals across different scenarios.

This evidence is presented in terms of the statistical model that represents preferences and is estimated on the basis of the observed choices. Coefficients of the model represent the average weighting of each unit/individual attribute (and associated level of that attribute) and thus are useful for describing (1) the relative importance of each attribute to respondents relative to a baseline and (2) the willingness of respondents to trade off one attribute for another.[8]

It is important to note that stated preference methods do not require respondents to actually make the types of choices represented in the experiment in the real world. In fact, were such data available, it would be preferred as revealed behavior and inference could be made on that basis. Lacking such data, however, a stated preference exercise that relies on hypothetical questions can provide some insight into the willingness of respondents to trade one attribute for another. However, we acknowledge that such methods are subject to potential hypothetical bias.

---

[8] If the equation is interpreted as an indirect utility function and is linear in the coefficients, then the coefficients can be interpreted as estimated marginal (relative) utility.

Organization of the Report

This chapter has described the background to this problem and our research approach. The rest of the report is organized as follows:

- Chapter Two explains the choice experiment methodology, provides an example of how to interpret model results, discusses the analysis of qualitative data used in the analysis, and explains how the findings were synthesized.
- Chapter Three explains the elicitation and analysis of data from the maneuver unit surveys in greater detail.
- Chapter Four explains the elicitation and analysis of data from the enabler units analyzed.
- Chapter Five explains the elicitation and analysis of data from individual surveys.
- Chapter Six summarizes the study's findings and provides overarching policy recommendations.

Appendix A presents the survey instruments used to collect data for this report. Appendix B briefly presents the general statistical model used to analyze the data.

CHAPTER TWO

The Choice Experiment Methodology

This chapter explains the choice experiment method employed in this research. It describes how the study team selected unit types for analysis, developed and validated survey instruments, and interpreted the results. The essence of this method was to infer the relative importance of various unit characteristics—including but not limited to Army component—in determining operational effectiveness from the choices experienced professionals made in a series of scenarios. Choices differed with respect to a number of characteristics, including unit type, previous in-country experience, training, and the proficiency of the unit's leadership teams, among others. Scenarios differed in terms of the nature of the challenge, the associated threat level, and the impact on risk to higher-echelon plans.

Overview

To evaluate the relative importance of characteristics of a unit or individual, we used stated preference method termed a "choice experiment." The general structure was for a respondent to choose the preferred alternative from a set, with different alternatives described by differing characteristics. To estimate preferences at varying levels of analysis, surveys were developed for BCTs, maneuver battalions, engineering battalions and companies, military police (MP) battalions, and individuals.

For each level of analysis, several scenarios were developed that described the role of a unit or individual in a given situation. The scenarios were described to respondents in detail (see Appendix A). Within a scenario, each survey presented a decision-maker with a series of tasks that involved choosing a hypothetical unit or individual that is perceived as the most attractive for a given scenario from among a set of three alternatives.

Each alternative was described by a set of characteristics. For example, for a maneuver battalion, the characteristics included the component, the level of postmobilization training (for ARNG units), the amount and/or type of in-country experience, and a representation of the opportunity costs—in terms of operational risks to the overall mission or continuity in the assigned area of operations—of choosing a particular unit. The set of characteristics varied by level of analysis (and, in some cases, across scenarios), but all included a representation of service component. The research team identified relevant characteristics for each unit type based on interviews with practitioners who had relevant experience. Included characteristics were initially made on the basis of the research team's experience and tested via focus groups. The characteristics were described to respondents in detail (see Appendix A).

While each alternative included the same characteristics, the levels of at least one characteristic differed between alternatives. This ensured that the alternatives were unique. The respondent was then asked to choose exactly one of the three alternatives that would be preferred for the described job or mission, assuming that the only information available is the level of each characteristic. Each choice is termed a "choice occasion." An example is provided in Table 2.1, with alternatives "Unit A," "Unit B," and "Unit C," and characteristics "Continuity," "Post-Mobilization Training," "Component," and "In-Country Experience."

Each scenario involved five choice occasions per respondent. The levels of the characteristics that appeared in each choice occasion were determined through the use of an experimental design constructed to ensure that there is enough information to estimate the chosen statistical model and analyze trade-offs between characteristics.[1] Statistical models were scenario-specific.

Table 2.1
Sample Choice Occasion for Unit-Level Analysis

For scenario 1, if you had to pick one of these three units, which would you select?

	Unit A	Unit B	Unit C
Continuity	High: available until relieved (> 9 months)	None: available for 3 months with no replacement for 3 months	Moderate: available for 6 months with immediate replacement
Post-Mobilization Training	High: 6 months post-mobilization, Mission Rehearsal Exercise (MRE) at Combat Training Center (CTC)	Not available: Regular Army unit	Moderate: 3 months post-mobilization training
Component	ARNG	Regular Army	ARNG
In-Country Experience	None	Counterinsurgency: 3 months	Security: 3 months

Choose ONLY one unit: _____

---

[1] Experimental designs for this study were constructed using the SAS software package. For more information about the use of experimental designs in choice experiments, see David A. Hensher, John M. Rose, and William H. Greene, *Applied Choice Analysis: A Primer*, Cambridge, UK: Cambridge University Press, 2005; and Barbara

Selecting Unit Types for Analysis

In conjunction with the sponsor, the research team focused its analysis on unit types that met two criteria. Unit types considered had been deployed enough that experience in Operations Iraqi Freedom and Enduring Freedom would have provided many potential respondents with a basis for comparing Regular Army, ARNG, and USAR unit performance, and conclusions about those unit types would be relevant to future Army policy decisions.

In consultation with the sponsor, we identified five different unit types for analysis. Given the importance of the question about how much the Army could rely on ARNG maneuver battalions to source contingency demands, it seemed obvious that we should include them in our analysis. Because maneuver battalions are organic elements of BCTs, we included them as well. Similarly, the Army's reliance on ARNG formations of various types to provide security forces, a role doctrine assigns to MPs, and the prominence of MP companies in the deployment data, indicated that we should include MP companies as a test of the degree to which the Army could continue to rely on ARNG and USAR security forces in the future. Aside from maneuver forces, engineers' route clearance mission may have been the most demanding and dangerous role in the context of Operations Iraqi Freedom and Enduring Freedom, leading us to analyze respondents' assessments of engineer battalions and companies. We initially included aviation battalions, but later dropped them when feedback from focus groups indicated that a stated preference survey for such units would require too many attributes for each choice occasion and thus prove too long and complex to administer.

Developing and Validating Survey Instruments

This section describes our general approach to developing and validating survey instruments; we will provide more information on the development of specific surveys in the following chapters. In broad outline, however, we developed separate surveys for the following unit types:

- *BCTs* are the Army's primary maneuver forces. The term *maneuver* indicates that they are responsible for conducting combined arms operations in a given area of operations. During the wars in Iraq and Afghanistan, they comprised about 4,000 soldiers, organized into several different types of battalions (e.g., reconnaissance, maneuver, fires, and sustainment).
- *Maneuver battalions* are part of BCTs and are responsible for conducting combined arms operations within a given—albeit smaller—area of operations. Maneuver battalions generally comprise companies of a single unit type (e.g., tank, infantry), although battalion headquarters are responsible for integrating

J. Kanninen, "Optimal Design for Multinomial Choice Experiments," *Journal of Marketing Research*, Vol. 39, No. 2, February, 2002, pp. 214–227.

effects provided by other arms and services. Maneuver battalions generally have around 700 soldiers.

- *Engineer battalions* are enabler units. Under the Army's modular organization, an engineer battalion is not a single unit containing several companies, but rather a command-and-control (C2) headquarters responsible for supporting, synchronizing, and integrating the efforts of attached engineer companies. The type of engineer companies attached can differ according to the mission and operational situation, but generally include horizontal construction, vertical construction, and route clearance in a counterinsurgency scenario.
- *Engineer companies.* We limited this analysis to route clearance companies. Route clearance companies are responsible for locating and neutralizing improvised explosive devices (IEDs) and other obstacles emplaced along the routes used by other U.S. and multinational forces and civilian traffic. They generally comprise fewer than 200 soldiers.
- *MP companies* provide law enforcement support, as well as other capabilities. For the purposes of this analysis, however, we focused on another key MP mission: route security. Like engineer battalions and companies, MP companies are an enabler formation.

We picked these unit types because mobilization and deployment data indicated that enough ARNG and USAR units of these types had been deployed such that there was a strong possibility that potential respondents would have been able to observe their operational performance.

Each survey consists of one or more scenarios. The study team generally developed scenarios by synthesizing information about specific historical events with observations from individuals we interviewed on the subject. For each scenario, respondents were presented with a number of options that varied along key dimensions identified through interviews with individuals who had experience working with reserve forces.

Survey instruments were then initially validated by military officers serving as fellows at RAND Arroyo Center recruited by the research team. Subsequently, hour-long focus groups led by members of the research team were held at the Army War College in late September 2013. Participants were recruited by officials at the school on the basis of their backgrounds, with two to five individuals per survey. Individuals were asked to complete a paper version of the survey and to make note of anything that was not clear, but to refrain from asking clarifying questions until everyone was finished. The group leaders then engaged participants in a semistructured discussion focused on specific elements of the survey, including instructions, clarity of descriptions of scenarios and attributes, relevance of the attributes and their levels to their choices, potential dominance of attributes, survey fatigue, and an open-ended discussion. Another member of the research team took notes about this discussion. Marked copies of the survey were collected from participants. Following the focus groups, the research team analyzed the comments of the focus group participants and refined the surveys.

Sample Population

We sought to elicit and analyze the views of officers whose experience provided a basis on which to assess performance of units and individuals of interest. To that end, we used Army Knowledge Online's mass email procedures to contact the approximately 2,400 Regular Army, ARNG, and USAR colonels in the following branches: infantry, armor, field artillery, special forces, engineers, military police, and logistics. An invitation email (under the cover of the director of the Army Quadrennial Defense Review Office) was sent with a link to the online survey (hosted by RAND's Survey Research Group) on June 5, 2014, with a follow-up reminder on June 17, 2014. No incentives were provided for response. We also contacted 412 Regular Army, ARNG, and USAR general officers from one of the foregoing basic branches via email on July 9, 2014.[2] The survey was officially closed in March 2015. From that group of approximately 2,400 potential respondents, 217 completed one of the surveys we developed, a response rate of slightly less than 10 percent.[3]

Statistically, we cannot be certain that our sample is representative of the larger population from which it is drawn, and, for some surveys, the number of respondents is small. The latter problem is, however, reflected in the standard errors of the estimated coefficients, which will tend to increase as the number of observations decreases. The former problem affects the ability to make credible inferences about the population of all Army colonels and general officers. If preferences of nonrespondents are substantially different than those in our sample, then our results will be contaminated with self-selection bias, and conclusions about the population from our models could be misleading. However, we have neither evidence nor expectations that this is the case.[4]

Our sample population had the kind and amount of experience necessary to speak authoritatively, however. Table 2.2 breaks down respondents by the most responsible position respondents held when deployed and their average months deployed. The sample population averaged over two years of deployed operational experience, although Regular Army respondents averaged almost 75 percent more time deployed. There were probably significant differences in the nature of their operational experiences as well. For example, Regular Army combat brigades deployed to Iraq and Afghanistan predominantly conducted close combat operations or conducted counterinsurgency missions. In contrast, other RAND research has indicated that only about nine of the 45 National Guard brigades deployed to Iraq and Afghanistan between

---

[2] We say "from" because general officers nominally have no branch.

[3] It was not possible to use state-of-the-art survey administration best-practices (e.g., the Dillman Tailored Design method) with this population, given the administration through the Army Knowledge Online system.

[4] We thus maintain the assumption that nonrespondents are missing at random.

Table 2.2
Sample Distribution by Most Responsible Position Held When Deployed and Average Months Deployed

	Active		Guard		Reserve	
	Count	Average Months Deployed	Count	Average Months Deployed	Count	Average Months Deployed
Commander/deputy commander at division or higher echelon	7	37.4	6	16.0	2	22.5
Chief of staff at division or higher echelon	4	44.3	3	23.0	1	12.0
Operations officer at division or higher echelon	3	33.3	3	10.7	2	33.0
Brigade combat team commander/deputy commander	7	30.3	11	23.6		
Special forces group commander			1	40.0		
Other brigade-level command	7	38.3	7	20.1	6	20.5
Maneuver/fires battalion commander	13	34.8	10	24.1		
Other battalion-level command	12	31.7	7	14.3	2	23.0
Transition team commander	1	40.0	9	21.3	5	19.4
Force management staff officer					2	8.0
Staff officer at division or higher echelon	13	34.5	13	19.0	16	30.2
Brigade/battalion-level staff officer	4	25.5	12	16.6	7	13.0
Other	2	58.0	12	15.1	7	13.9
Total	73	34.6	94	19.1	50	21.1

2003 and 2010 performed in a counterinsurgency role. The remainder performed security force (SECFOR) missions or assisted in training and advising indigenous forces.[5]

---

[5] Ellen M. Pint, Matthew W. Lewis, Thomas F. Lippiatt, Philip Hall-Partyka, Jonathan P. Wong, and Tony Puharic, *Active Component Responsibility in Reserve Component Pre- and Postmobilization Training*, Santa Monica, Calif.: RAND Corporation, RR-738-A, 2015, pp. 49–51.

Respondents took different surveys depending on their branch and the most responsible position held when deployed according to biographical questions posed at the survey's outset. Although we do not have good data about the actual time that respondents took to answer the surveys, focus group participants generally completed the survey (and provided notes) in paper form within 30 minutes. Table 2.3 describes how respondents were aligned with surveys. For example, commanders and deputy commanders at the division level or higher and BCT commanders and operations officers at the division level or higher were supposed to take the BCT survey. In subsequent chapters, we will describe the sample that responded to each survey.

Each survey had a fixed design; that is, no randomization occurred within a choice occasion, between choice occasions, or between scenarios. This decision was made primarily for administrative convenience. However, it should be noted that this decision could result in anchoring bias, in which survey fatigue causes a respondent to respond based on simple rules that do not fully reflect their preferences. Were this the case, then the estimated model coefficients would be biased. We have no particular evidence of anchoring bias across the surveys, as focus group participants did not report any losses of concentration (after asking specifically about this issue), and results across scenarios generally match our a priori expectations.

Developing the Statistical Models

We used a conditional (sometimes called a multinomial) logit model to analyze the stated choices. This model is appropriate because the dependent variable is discrete; that is, it indicates which of the three offered choices per occasion a particular respondent chose. The explanatory variables in each model consist of the unit or individual

Table 2.3
Alignment of Surveys with Respondents

Survey	Respondents
BCT	Commanders and deputy commanders at the division level or higher; operations officers at the division level or higher; BCT commanders; special forces group commanders
Maneuver battalion	BCT deputy commanders; maneuver or fires battalion commanders; other brigade- or battalion-level commanders from infantry, armor, field artillery, or special forces branches; force management staff officers
Engineer battalion and company	All engineer officers
Military police	All military police officers and brigade- and battalion-level commanders from logistics branches
Individual	Chiefs of staff at division level and higher; all other categories not previously mentioned

characteristics and levels expressed as indicator variables.[6] In brief, the model fits coefficients to best predict the observed choices using the characteristics associated with each potential choice in a set as the data. Appendix B provides additional information.

In the models estimated herein, the equation that describes the average preferences of respondents across the sample take the form:

$$\hat{V} = \hat{\beta}_1 x_1 + \hat{\beta}_2 x_2 + \ldots + \hat{\beta}_K x_K ,$$

where x_k is an indicator for each characteristic and level presented to the respondent across all choices in a scenario, and $\hat{\beta}_K$ are the estimated weights associated with each one.[7]

Interpreting the Statistical Models

The estimated weighting coefficients $\hat{\beta} = (\hat{\beta}_1, \hat{\beta}_2, \ldots, \hat{\beta}_K)$ provide a measure of the importance of each characteristic relative to the others. As such, these coefficients can be used to identify preferred alternatives and to explore the trade-offs between attribute levels.

For example, suppose an alternative comprises two characteristics: training (characteristic x_1, which can take on levels of "three months" or "six months"; and in-country experience (characteristic x_2), which can take on levels of "three months security" and "three months counterinsurgency." Define the reference category to be three months training and three months security experience,[8] with the variable $x_1 = 1$ indicating six months training and the variable $x_2 = 1$ indicating three months counterinsurgency experience. The model then takes the form $V = \beta_1 x_1 + \beta_2 x_2$.

Suppose a model is estimated such that

$$\hat{V} = 2x_1 + 1.5x_2 ,$$

where $\hat{\beta}_1 = 2, \hat{\beta}_2 = 1.5$, and a "hat" indicates an estimate. The results for this model (assuming small, randomly chosen standard errors) are reported in Table 2.4.

---

[6] So, for example, an in-country experience variable with three levels (none, three months, and six months) will enter the model with one indicator variable that equals one if in-country experience for the unit is three months and zero otherwise and another indicator variable that equals one if in-country experience for the unit is six months and zero otherwise. In this case, in-country experience of zero months is the baseline category. In principle, one could add socioeconomic characteristics (such as years deployed); however, testing for heterogeneity in preferences other than respondent component was not a key research question in this study. As such, the models represent average preferences across the sample.

[7] As described in Appendix B, the estimated model that describes the probability of making a particular choice is nonlinear.

[8] Because the index is an ordinal, rather than cardinal, measure, a reference category is required for model identification. The index value for this case is arbitrarily set to zero by construction.

Table 2.4
Example Model Output

Variables	Coefficient Estimates
Training—six months	2.000***
	(0.289)
Experience—counterinsurgency	1.500***
	(0.230)

NOTES: Standard errors in parentheses. *** $p < 0.01$, ** $p < 0.05$, * $p < 0.1$. Baseline category is unit with three months training and three months security experience.

The estimated underlying scale values for each of the possible four alternatives for this example are given in Table 2.5.

In this example, a unit with six months of predeployment training and three months of in-country counterinsurgency experience is the preferred option among the four unit alternatives, as this unit has the highest scale value.[9] Similarly, when comparing any two units, the unit with the highest scale value is preferred on average. As such, a unit with three months of predeployment training and three months of in-country counterinsurgency experience is preferred to a unit with three months of predeployment training and three months of in-country security experience (index values of 1.5 versus 0, respectively).

Furthermore, the values of the coefficients provide information about the trade-offs between attribute levels. All else equal, the average respondent would prefer a unit with six months of predeployment training and security experience to a unit with three months of training and counterinsurgency experience (scale values of 2 to 1.5, respectively). This implies that the three additional months of predeployment training is more valuable to the average respondent than in-country counterinsurgency experience

Table 2.5
Scale Values for Training/Experience Example

Example Alternative	Index Value
Three months training, three months security experience (baseline category)	2(0) + 1.5(0) = 0
Six months training, three months security experience	2(1) + 1.5(0) = 2
Three months training, three months counterinsurgency experience	2(0) + 1.5(1) = 1.5
Six months training, three months counterinsurgency experience	2(1) + 1.5(1) = 3.5

---

[9]   Note that this also implies the highest probability of being chosen.

(relative to security experience), or equivalently, that the ratio $\beta_1 / \beta_2 = 1.33$ is greater than one. Note that this is true despite the fact that both longer predeployment training and in-country counterinsurgency experience are preferred overall (as given by the positive and statistically significant coefficients on each variable). Using this logic, the value of any characteristic included in a conditional logit model can be expressed in terms of any one of the other characteristics.

Indicator variables can also be used to estimate differences in preferences across individuals in different groups. We use this technique to test for differences between Regular Army, ARNG, and USAR respondents. To illustrate, let d be a variable that is equal to one if the respondent is a Regular Army officer (0 otherwise). We can statistically test for differences in preferences between these two groups by estimating the model $V = \beta_1 x_1 + \beta_2 x_2 + \beta_3 x_1 d + \beta_4 x_2 d$, and testing if the estimates for β_3 and/or β_4 are different from zero. If they are, we conclude that there are preference differences between the two groups (Regular Army and other).

In the results presented in the next chapter, we used three scenarios representing different threat levels to test if results changed with the threat level or were robust to such changes. Readers should note that comparison of absolute values of coefficients for this type of model is not necessarily meaningful because the probability of a particular choice (which is what is being modeled) depends on all of the coefficients in the model. In other words, if $\beta_1 = 1$ for the low-threat scenario and $\beta_1 = 2$ in the high-threat scenario, this does not necessarily mean that the underlying attribute (x_1) is more important in the high-threat scenario.

To see why, consider the case where $\beta_2 = 1$ in the low-threat scenario and $\beta_2 = 3$ in the high-threat scenario. In terms of the underlying second attribute (x_2), x_1 is actually less valuable in the high-threat scenario (with a ratio of 2:3) than in the low-threat scenario (with a ratio of 1/1).

Analysis of Qualitative Data

We also analyzed qualitative data provided by respondents in the course of the survey. This analysis provides additional insights into the characteristics that mattered to respondents when selecting each unit. This analysis is based on open-ended questions asked at the end of each choice experiment. Specifically, the respondents were asked: "In a couple of sentences, please describe any other unit characteristics that would be important to your decision." The question was intended to collect information about any additional attributes that may have mattered to respondents when they were making their choice even though they were asked to assume that units were identical along all other qualities not mentioned in the choice experiment.

Our analysis focused on the following questions: Do these qualitative responses contain any information that may question the validity of the quantitative results? Did the respondents have sufficient information to engage in the exercise in a meaningful way? What additional assumptions did respondents make when answering survey question? For example, a high number of responses indicating that the scenario was not clear or lacked critical information about the mission would have been a serious concern for the validity of our results.

After reviewing each of the responses, we organized them into one of the three categories: (1) comments reiterating the importance of attributes already listed in the choice experiment, (2) comments mentioning attributes not listed in the choice experiment, and (3) comments on the clarity or plausibility of the scenario. Table 2.6 provides examples from each survey that were included in each of the three categories.

Table 2.6
Examples of Qualitative Responses Provided

	Redundant	New Attribute	Scenario
BCTs	"High-risk environment makes employing inexperienced units or ones trained for a different mission less suitable for this task." "Criteria that matters to me: (1) in-country experience, (2) type of unit (in this scenario, ABCT [armored brigade combat team] over the infantry BCT [IBCT]), and (3) then component."	"Leadership (command and senior noncommissioned officers) make a difference and must be considered." "Performance of those units who have been in theater. In-theater missions were briefly outlined, so that helps, but still may not be best suited to assume the outgoing IBCT mission set."	"It would be helpful to know more about the importance of the planned upcoming offensive that would be delayed. Also I have been making an assumption that a BCT that has been scheduled to rotate in to replace the outgoing one trained specifically for that mission and terrain."
Battalions	"Collective training and in-country experience are still factors, however, length of deployment time is now most critical for building governance and host-nation special forces."	"The Battalion Commanders combat experience." "For those with in-country experience: assessment of performance and leadership. For all: unit counterinsurgency skills inventory—other than military occupational specialty skills available in the organization—language, civilian police, power generation, etc."	"This scenario moves closer to decisive action than the previous two and decreases the weight given to the duration of availability."

Table 2.6—Continued

	Redundant	New Attribute	Scenario
Military police	"Validation of tasks is important so the unit knows its job and can relay those skills to the host nation counterpart." "Unit validation and manning, there was no difference in my mind when referencing U.S. Army Reserve versus active component."	"Language skills of soldiers" "The experience of the leadership" "Time available to unit to conduct mission. More time is desirable."	"Cost row was confusing: Why would we bring two companies over to replace one? In four of the five scenarios, the costs of two companies were presented as unit is deploying to replace the selected unit. This was not clear." "The tempo of operations in the other sector, political implications, and the time left in country for the unit that may move within sector. Can I send the newly deploying unit to the new sector?"
Engineers	"Company leadership is number one; unit experience is also important. Training/ certification can be achieved while in county."	"Time in country experience average physical fitness test scores/Medical Protection System (MEDPROS) qualification (shows thoroughness of company commander) equipment maintenance—OR [operational readiness] rate supply accountability— on hand rate (good OR rates and supply rates indicate strong units)."	"Note: The company scenarios have several conflicts—units can't have low experience and be remissioned from an active sector. Please relook this."
Individual	"Individual must have: prior deployment experience operational experience meet time requirements for vacancy Vice President of the small trucking company: He/ she seems to be a proven manager (maybe leader) at keeping several balls in the air in real time - current operations" "The person with the most time available. Civilian acquired skills are important, but longevity in position is more."	"Must have in-depth understanding of US Code and fiscal law. Must have experience developing business plans and creating economic networks. Must have ability to understand local business customs and practices and be culturally aware." "Skilled in recognizing corruption, theft and mismanagement."	"This survey is be skewed toward NG/USAR since most NG/USAR soldiers possess a civilian skill outside of their military duties. AC soldiers may possess similar skills; unfortunately this survey does not show that experience and as a result they are at a disadvantage in this survey showing Has no relevant civilian acquired skills."

For each of the surveys, we examined responses in categories 2 and 3 more closely. We paid less attention to category 1 because responses included in this category did not provide any additional information about the choices that respondents recorded during the multiple-choice section of the exercise. Analyzing these responses was not going to add any new information to what we already know from the statistical analysis. We summarize the results of our qualitative analysis below for each survey.

Synthesizing Findings

The models developed from our statistical analysis told us how our respondents weighed selected attributes in a specific, yet hypothetical, situation. It was necessary to extrapolate from these analyses to develop overarching findings from which to derive policy recommendations. We synthesized these overarching findings from the totality of the available data and analysis. The resulting narrative tends to emphasize describing Regular Army officers' perspectives and preferences. To the extent that component status mattered, it mattered to them. Ergo, assessing the strength of this preference and identifying the conditions that could overcome it were of greatest interest. Since we could not resolve the different perspectives between Regular Army, ARNG, and USAR officers, we present all of them in the body of this report.

CHAPTER THREE

Maneuver Units

The analysis in this chapter assesses the degree to which the Army should rely on ARNG maneuver units to meet operational requirements for maneuver forces in counterinsurgency operations and the conditions under which it should do so. For the purposes of this analysis, maneuver forces included BCTs and maneuver battalions. The chapter describes the results from our statistical analysis of survey responses.

In each case (BCTs and maneuver battalions), we developed three scenarios corresponding to varying "threat levels." In so doing, we attempted to describe realistic scenarios, loosely based on operational experience, that were informative, case specific, and meaningfully different across scenarios. To do so, and in conjunction with feedback from the validation exercises described in Chapter Two, the narratives for each scenario were similar but not identical. While this may introduce a source of extraneous variation (other than the attributes and their values) when comparing across cases, it also provides the opportunity for robustness checks of results.

Similarly, because the stated preference exercise is specifically designed to create trade-offs within choice occasions, it was important to develop an attribute that represented the opportunity costs of a decision that would, in some way, be borne by the respondent (in terms of potential outcomes about which he or she cared). Given the novelty of the context, it was not obvious what this attribute should be. After careful consideration, the research team decided to represent opportunity costs in two different ways: "overall operational risk" for BCTs and "continuity" for maneuver battalions. While this choice also introduced a source of variation across cases, it again provided an opportunity to check robustness and survey validity (in terms of a priori expectations about different levels of these variables) and provided the opportunity to test if one or both of these concepts is seen as valid as a cost measure in this type of context.

In brief, the analysis indicates that Regular Army respondents believed Regular Army maneuver units were significantly more capable than their ARNG counterparts, all other things being equal, while ARNG respondents did not. In the scenarios, as in real life, all things were not equal, and even Regular Army respondents found ARNG units preferable under different sets of conditions in low- and moderate-threat environments. More specifically, Regular Army officers generally preferred employing ARNG units to incurring significant opportunity costs, denominated either in terms of operational risk to the commander's overall plan or in terms of sacrificing continuity in the area of operations.[1] Regular Army respondents generally felt that Regular Army units were a better choice for high-threat environments.

As a reminder, the relative importance of various factors within a given scenario is represented by their coefficients presented in tabular form. While sophisticated statistical analysis was required to develop the models represented by these coefficients, interpreting them simply requires comparing their *relative* numerical values. Readers should not, however, compare *absolute* coefficient values across scenarios.

BCT Analysis

The BCT is the Army's primary maneuver formation. Throughout much of the wars in Iraq and Afghanistan—beginning in late 2004 and continuing through 2013—the standard BCT comprised six battalions of different types and consisted of approximately 4,000 soldiers. BCTs were responsible for integrating all operations and effects provided by organic, attached, and supporting units within a given area of operations. The ability of the BCT commander and staff to integrate and synchronize these different effects is considered a critical factor in the unit's overall effectiveness.[2]

To assess respondents' preferences for different unit characteristics, presumably representing their assessments of those characteristics relative importance in determining operational effectiveness, the research team developed three BCT scenarios. Each scenario represented three levels of threat. Using three threat levels allowed the research team to determine the degree to which differing environments affected respondents' valuation of unit characteristics. Scenarios were loosely based on actual historical examples from Operation Iraqi Freedom. For the low- and moderate-threat scenarios, the research team used examples in which ARNG units had actually been employed. The scenarios did not actually identify the historical examples from which they were drawn.

Unit Characteristics

As noted, the research team identified salient characteristics through interviews and reviews of the available literature. BCT characteristics and potential levels for all three scenarios included:

- **Overall operational risk** (three levels): Operational risk was defined in terms of the additional risk that selecting a particular option will incur in terms of

---

[1] We varied the notion of cost across surveys to test which measure would resonate with respondents.

[2] For a more detailed description of BCTs as constituted in this period, see Headquarters, Department of the Army, "The Brigade Combat Team," Washington, D.C., Field Manual (FM) 3-90.6, 2006.

the commander's overall plan, particularly in terms of impact on a forthcoming offensive to reclaim a key province from the insurgency. In making their choice, respondents had to balance the risk in the area of operations in question against the risk to the corps commander's overall plan.

- *No additional operational risk:* This unit was already scheduled to replace the outgoing BCT. Selecting this particular unit would not increase risk to the commander's plan.
- *Moderate operational risk:* This unit would have to be diverted from a lower-priority mission. It was possible, but not assured, that forces would be able to mitigate the risks in the mission from which the other unit might be diverted.
- *High operational risk:* This unit was intended to participate in the forthcoming offensive. Selecting this unit to perform the mission described in the scenario means that it would not be available in the forthcoming offensive, which will would to be deferred for several months as a result.

- **In-Country Experience** (three levels): All units were available for the mission for up to six months, but had varying degrees of in-country experience.
  - *None:* BCTs with no prior in-country experience.
  - *Security:* BCTs with three months of in-country experience providing installation and route security.
  - *Counterinsurgency:* BCTs with three months of in-country experience conducting counterinsurgency operations.

- **Postmobilization Training** (two levels, ARNG): For all ARNG units, the information about training conducted as a unit to prepare for deployment was quantified. It included training conducted while mobilized under Title 32 authority immediately prior to mobilization under Title 10, as well as training conducted after Title 10 mobilization.
  - *Moderate:* BCT had completed three months of postmobilization training prior to arriving in the theater. In recent overseas contingency operations, three months of training was considered sufficient to prepare units for SECFOR operations conducted away from forward operating bases (FOBs).
  - *High*: BCT had completed six months of postmobilization training, culminating in an MRE at a CTC.

- Force Structure Component (two levels):
  - ARNG
  - Regular Army.

- **Unit type** (two levels): Units were either IBCTs or heavy BCTs (HBCTs), also known as armored BCTs (ABCTs).
  - *ABCT:* Also known as HBCT
  - *IBCT.*

Appendix A provides the survey instruments for the BCT level, including the full description of the scenarios and unit characteristics.

Respondents

Either 34 or 35 individuals answered each of the BCT survey questions in each of the three scenarios.[3] Respondents brought a significant amount of relevant experience to the exercise: 17 general officers and 18 colonels participated in the survey. Table 3.1 shows the breakdown of these respondents by the count of most responsible position held and the average number of months deployed.

BCT Statistical Results

Table 3.2 presents the results for BCTs for each scenario. A discussion of results for each scenario environment follows. Variables are listed in the left-hand column. The next six columns list the coefficients for those variables in each scenario; asterisks indicate the variables are statistically significant in the model in question. Two models are presented for each scenario. Model 1 in each scenario indicates the results when all respondents are combined. Because analysis indicated that Regular Army and ARNG respondents made different choices, we analyzed their responses separately, with the results indicated in the columns labeled "Model 2."

Table 3.1
BCT Survey Sample Distribution by Most Responsible Position

Position	Regular Army		ARNG	
	Number	**Average Months Deployed**	**Number**	**Average Months Deployed**
Commander/deputy commander at division level or higher	7	37.4	8	17.6
Operations officer at division level or higher	3	33.3	4	17
BCT commander	6	29.4	6	24.8
Special forces group commander			1	40

---

[3] There were 34 responses for scenario 1, choice 3, and scenario 2, choice 4, with different respondents choosing not to answer those questions. The reason(s) for the missing responses is not clear. As such, all available information (i.e., the four answers provided by these respondents) was used in estimation.

Table 3.2
Choice Experiment Results, BCT Units

Variables	Low-Threat Environment		Moderate-Threat Environment		High-Threat Environment	
	Model 1	Model 2	Model 1	Model 2	Model 1	Model 2
Regular Army unit	1.081***	0.006	0.132	-0.374	0.685**	0.126
	(0.412)	(0.388)	(0.245)	(0.260)	(0.305)	(0.377)
Train—high (ARNG only)	0.414**	0.428**	0.360**	0.370**	0.205	0.255
	(0.206)	(0.208)	(0.161)	(0.163)	(0.452)	(0.459)
Associated levels of operational risk—moderate	2.353***	2.625***	0.861**	0.914**	0.704**	0.764**
	(0.485)	(0.523)	(0.397)	(0.388)	(0.308)	(0.313)
Associated levels of operational risk—low	3.606***	4.120***	1.842***	1.944***	1.527***	1.629***
	(0.548)	(0.619)	(0.413)	(0.413)	(0.388)	(0.408)
Experience—security	0.844***	0.821**	0.565**	0.545*	0.828***	0.848***
	(0.308)	(0.327)	(0.272)	(0.281)	(0.300)	(0.302)
Experience—counterinsurgency	1.370***	1.615***	1.651***	1.702***	1.387***	1.479***
	(0.261)	(0.272)	(0.269)	(0.264)	(0.388)	(0.384)
Combined arms (heavy)	−0.216	−0.233	−0.0654	−0.0755	0.685***	0.700***
	(0.241)	(0.245)	(0.217)	(0.217)	(0.259)	(0.259)
Regular Army respondent *Regular Army unit	—	2.353***	—	1.107***	—	1.277**
		(0.657)		(0.430)		(0.535)
Number of observations	35	35	35	35	35	35
Number of choices	522	522	522	522	525	525

NOTES: Multinomial logit model results with clustered (by individual) standard errors in parentheses. Model 1 pools all respondents. Model 2 is final restricted model (zero coefficient restrictions imposed) allowing for different coefficients by respondent component. *** $p < 0.01$, ** $p < 0.05$, * $p < 0.1$. Baseline category is ARNG IBCT with moderate postmobilization training, high associated levels of operational risk, and no experience. Training: High variable does not apply to Regular Army units. Absolute values of coefficients are not comparable across scenarios.

The complexity of the different situations and the interplay of different variables within those situations make generalizations difficult. A few things stand out, however. First, the results in Table 3.2 show that the single most important factor for all respondents within a given model was keeping the overall level of operational risk as low as possible. Second, as the level of risk increased, respondents accorded more importance to prior experience conducting counterinsurgency operations. Third, as risk increased, the weight that Regular Army respondents placed on using Regular Army maneuver units also increased. We will explore the results of the different analyses in greater depth in the following sections.

Low-Threat Scenario Results

Scenario 1 is the "low-threat" environment, though this language was not used in the survey. Instead, the description corresponded roughly to historical experience deemed to be lower threat. In this analysis, Regular Army officers chose to employ Regular Army units, all other things being equal. In the context of the choice experiment, however, maintaining low levels of operational risk was even more important, by a factor of about 60 percent. We can see that, by comparing the value of maintaining different levels of operational risk—represented by the coefficients shown—with the value to Regular officers of employing a Regular Army unit. Even maintaining moderate levels of operational risk through the choice of unit was about as important to Regular Army respondents as being able to employ a Regular Army unit.

Respondents were told that they were the deputy operations officer for a multinational corps conducting counterinsurgency operations in a medium-sized developing nation. Their task was to develop a recommendation for replacing an IBCT that is scheduled to redeploy in two months.

The operational environment was described as one in which the departing IBCT's area of operations consisted mostly of urban and close terrain, with two medium-sized cities with populations of 300,000 and 500,000. According to the scenario, the BCT was to be deployed to an area of operations where a U.S. division, employing the theater reserve, had recently crushed an insurgent uprising in one of the cities in the area of operation. The operation inflicted heavy losses on the insurgents, but insurgent networks still remained entrenched. The residual threat level was described as sporadic and of variable intensity. Respondents were informed that the corps intelligence officer believed that the IBCT's offensives had significantly disrupted the insurgency in the province. At the time of the decision, most attacks on coalition forces took the form of IED and rocket attacks, though the enemy continued a campaign of assassination and intimidation against host-nation forces. The incoming unit was to consolidate the outgoing unit's success and transition to stability operations, with emphasis on developing host-nation security forces and supporting economic development. In this scenario, respondents were to select a BCT to conduct counterinsurgency operations in a mostly urban area of operations in which insurgent activity was relatively sporadic and ineffective.[4]

---

[4] This scenario was based on the situation in a Najaf province after the defeat of the two major Jaish al Mahdi uprisings in April and August of 2004. In early 2004, elements of the 155th Brigade, Mississippi Army National

Respondents were informed that the scenario descriptions and characteristics of the potential incoming units were the only distinguishing information available at the time of choice. They were to assume that all units were available for up to six months and were equipped at the same level, except as indicated by unit type (i.e., IBCT or ABCT).

The first data column in Table 3.2 (low-threat environment, Model 1) reports the results from this scenario using the complete sample. All characteristics enter the model as indicator (or "dummy") variables. Such variables take on a value of one if a characteristic takes the value indicated in the "variables" column and zero otherwise. The reference case (for which all indicator variables are assigned a value of zero) is an ARNG IBCT unit with moderate levels of training, high associated levels of operational risk, and no in-country experience.

The results for Model 1 show that, all else equal, respondents would choose to employ Regular Army units $\beta_{Active} = 1.081$. In other words, if all other variables associated with Regular Army and ARNG maneuver units were identical (including unit type, experience, and levels of operational risk), respondents would choose the Regular Army unit. Similarly, all else equal, respondents would prefer to employ units associated with lower levels of operational risk (relative to the baseline of "high") prefer counterinsurgency experience to security experience, and for ARNG maneuver units, prefer higher levels of training. These results are as expected. For this low-threat scenario, there is no evidence that respondents accorded any particular weight to unit type—IBCT or ABCT—because of the statistical insignificance of the associated coefficient.

The preferences implied in the table suggest that, on average, respondents are willing to trade their preferred component attribute for lower levels of operational risk and/or higher levels of actual operational experience, e.g., three months of in-country counterinsurgency experience. That is, if faced with a choice between a Regular Army unit needed for a planned offensive—and whose diversion to the tactical situation in question would thus incur higher operational risk—and an ARNG unit whose choice would not affect the planned offensive, the average respondent would choose the ARNG maneuver unit. Similarly, if operational risk levels were identical between units, the Regular Army unit had no in-country experience, and the ARNG unit had three months of counterinsurgency experience, respondents would tend to employ the ARNG unit. However, this would no longer be the case if the Regular Army unit had three months of security experience. In addition, there is weak evidence that respondents would choose an ARNG unit with high levels of predeployment training plus three months of in-country security experience over a Regular Army unit with high levels of associated operational risk, all else equal. In short, respondents were willing to trade ARNG units for Regular Army units if such choices would either reduce disrup-

---

Guard were responsible for counterinsurgency operations; in Kevin Reeves, "155th BCT Hits the Ground Running," *Marine Corps News*, reprinted at GlobalSecurity.org, February 19, 2005.

tions to the overall campaign plan or increase the level of actual operational experience in the unit to be employed.

In terms of relative value, all else equal, respondents appeared to value Regular Army units with no prior in-country experience about 28 percent more than an ARNG BCT with three months of in-country experience conducting security operations.[5] Regular Army status was over 161 percent more important than nine months of postmobilization training for an ARNG unit, suggesting that Regular Army respondents did not judge that even extensive predeployment training could bring ARNG units to levels of effectiveness comparable with their Regular Army counterparts.

Said another way, Regular Army respondents valued Regular Army units highly, but some factors were even more important to them. Minimizing operational risk was more important to respondents than being able to employ a Regular Army unit. The coefficient associated with keeping operational risk low was 4.120 (shown in Table 3.2, low-threat environment, Model 2), 75 percent higher than the coefficient for getting a Regular Army unit, 2.353. That means that respondents felt it was almost twice as important to keep the overall level of operational risk low than it was to get a Regular Army unit. Even keeping risk moderate was slightly more important to Regular Army respondents, with a coefficient of 2.625. Other factors could compensate for Regular Army status. For example, Regular Army respondents valued an ARNG BCT with three months of prior in-country experience conducting counterinsurgency operations—as opposed to security operations—as having almost 80 percent of the value of a Regular Army unit. In short, while Regular Army respondents would rather be able to employ a Regular Army unit all else equal, they would much rather avoid disrupting the overall campaign plan and employing an ARNG unit to employing a Regular Army unit and disrupting the plan.

To test whether or not preferences differed between Regular Army and ARNG respondents, a model was estimated that allowed the coefficients of the choice model to vary between groups, as discussed in Chapter Two. Statistical tests were performed to test if the coefficients were statistically different between groups. Applying the implications of these tests to the expanded model results in the estimates shown in the second data column of Table 3.2 (Model 2).[6]

Results show that only Regular Army respondents accorded a higher weight to employing Regular Army units. This can be seen by the lack of significance on the *Regular Army Unit* variable (indicating that ARNG respondents did not tend to consider component status as a significant predictor of unit performance) and the sig-

---

[5]   Percentages are based on the ratio of the active coefficient to the other attribute that varies from the baseline.

[6]   Specifically, a Wald test that the seven new interaction variables associated with Regular Army personnel were equal to zero was rejected at a p value of 0.0000 (Chi-squared test statistic of 57.42 with 7 degrees of freedom), but a likelihood ratio test of all but the interaction term related to component was not rejected at a significance level of 0.12 (p value of test = 0.12, test statistic = 10.10, 6 degrees of freedom). As such, in the table, the six jointly insignificant coefficients were set to zero for Model 2.

nificance of the interaction term *Regular Army Respondent*Regular Army unit* (indicating that Regular Army respondents did consider component status to be a significant predictor of unit performance). In other words, service component was not a major driver of responses for ARNG respondents (or alternatively, the value of being a Regular Army versus ARNG unit is statistically equal to zero), but was for Regular Army respondents.[7] Results for the other unit characteristics were largely unchanged, at least qualitatively.

Moderate-Threat Scenario Results

Scenario 2 was the "moderate-threat" environment, though this was not indicated to respondents. The circumstances in the scenario corresponded to a situation in which an ARNG brigade had been employed. Results from this scenario were somewhat anomalous. While the scenario described is more challenging than the low-threat environment, Regular Army respondents did not weigh Regular Army unit status as heavily as they had in the lower-threat scenario. It also varies substantially from the findings with regard to the high-threat environment, in which Regular Army respondents considered Regular Army units more appropriate regardless of the associated level of operational risk. In this moderate-threat scenario, Regular Army respondents valued being able to employ Regular Army units slightly more than ARNG units, but the difference in attribute values is not especially large. Almost every other statistically significant factor outweighs component status in importance, however.

The role of the respondent was unchanged from the low-threat environment. However, the operational environment was changed to one in which an HBCT (ABCT) was being redeployed in two months, with an area of operations centered on the provincial capital, a city of about 500,000 inhabitants.

The scenario described the operational environment as "challenging," in that neither the host-nation government—supported by coalition forces—nor the insurgents hold sway. Respondents were told that the corps intelligence officer believed that the insurgent headquarters for the province was located in the city. Insurgents mostly employed IEDs with varying degrees of sophistication to attack U.S. troops, but could also conduct attacks by fire and small-scale ambushes. At the time of the decision, insurgent forces had seized several checkpoints manned by host-nation security forces. According to intelligence reports, insurgent activity and capability in the city had not yet peaked, but was trending upward.[8]

---

[7] Note that there is not a symmetric preference for ARNG units for ARNG respondents, which would be indicated by a negative and significant coefficient on *Regular Army unit*.

[8] This scenario was loosely modeled on the situation in Ramadi in 2005, when the 2nd Brigade, 28th Infantry Division (Pennsylvania Army National Guard) replaced the 2nd Brigade, 2nd Infantry Division, as described to a research team member by senior Marine Corps officers with oversight responsibility for the area of operations. See also John L. Gronski, Kurt Nielsen, and Alfred A. Smith, "2/28 BCT Goes to War," undated. Gronski commanded 2/28 BCT during this deployment.

The incoming unit mission was described as replacing the outgoing ABCT. The mission for the new unit was the same as for the outgoing unit; namely, to neutralize the insurgency in the area of operations, with a secondary mission to develop host-nation security forces. In the past, the outgoing BCT operated with its assigned armored vehicles, but generally operated either dismounted or in up-armored high-mobility multipurpose wheeled vehicles (HMMWVs) and mine-resistant ambush protected (MRAP) vehicles.

Respondents were again informed that the scenario descriptions and characteristics of the potential incoming units were the only distinguishing information available at the time of choice. They were to maintain the assumption that all units were available for up to six months and are equipped at the same level, except as indicated by unit type (i.e., IBCT or ABCT).

The third column of Table 3.2 (moderate-threat environment, Model 1) present the results from the choice experiment for this scenario using the complete sample with the same reference case as the low-threat scenario. While associated levels of operational risk and experience variables remained significant (and of the expected sign) at the 5-percent level, as in the low-threat scenario, the coefficient on *Regular Army unit* could not be distinguished from zero with the data available. As such, there was no average estimated preference across all respondents toward Regular Army BCTs for the moderate-threat environment.

Moreover, respondents appeared to accord no particular importance to unit type. The scenario described was significantly more challenging than that in the low-threat scenario, with respondents employing IEDs and occasionally small-unit attacks. Also, the unit being replaced was an ABCT, a unit with more mobility, protection, and firepower than an IBCT, although the unit typically employed HMMWVs and MRAPs to conduct operations. In spite of these factors, there was no statistical difference in the weight respondents accorded to IBCTs and ABCTs. It is unclear if this actually represents respondents' assessment of moderate risk contexts, or merely that the sample size was too small to uncover the difference.

Testing for equivalence in preferences between Regular Army and ARNG respondents in the moderate-threat scenario (fourth data column of Table 3.2, moderate-threat environment, Model 2) led to similar insights with respect to component as the low-threat scenario. On average across all respondents, there was no preference for Regular Army maneuver units, nor could a test of equivalent coefficients between Regular Army and ARNG respondents be rejected at the 90-percent level of confidence.[9] However, when estimating a model that imposes identical coefficients for all variables other than *Regular Army unit*, Regular Army respondents did appear more inclined to choose Regular Army BCTs for the moderate-threat environment. We conclude that

---

[9] Test of equivalence results in p value of 0.21, with a Chi-squared test statistic of 9.49 with 7 degrees of freedom.

Regular Army respondents tended to assess that Regular Army units are more appropriate for this scenario. ARNG respondents did not appear to share this assessment. Neither our survey nor other available data enable us to authoritatively explain the ARNG respondent perceptions.

We remind the reader that, in interpreting results across scenarios, relative (rather than absolute) values are the meaningful metric. That is, although the coefficient on the interaction term is larger for the low-threat environment than the moderate-threat environment, this does not mean that respondents valued the Regular Army maneuver unit more in the low-threat scenario. In fact, in terms of moderate operational risk (which is a proxy for overall opportunity cost), Regular Army respondents actually valued Regular Army component status of the unit *less*, as the ratio of these two coefficients is less in the low-threat scenario than in the moderate.

High-Threat Scenario Results

Scenario 3 was the "high-threat" environment, though this scenario did not explicitly characterize the scenario in these terms. The details of the situation corresponded to a situation at the beginning of the "surge" and in which Regular Army maneuver forces were challenged. In this situation, Regular Army respondents again placed a high value on Regular Army maneuver units. The role of the respondent was unchanged from the low- or moderate-threat environments. However, the operational environment was changed to one in which the threat was greater and the area of operations was more distributed. Respondents were told that the area of operations was dominated by insurgents, with lines of communication eventually tracing back to insurgent sanctuaries in other parts of the host-nation and neighboring countries. The area of operations was wide, combining several small- to medium-sized communities ranging in size from 25,000 to 100,000 people along major rivers and roads. Respondents were explicitly informed that the terrain seemed well suited to either an ABCT or a Stryker BCT (SBCT).

The threat level was described as one in which insurgents were capable of attacking U.S. forces with IEDs, rockets and mortars, and coordinated attacks using fire and maneuver. One attack was described as a platoon-sized insurgent element conducting a raid on a small U.S. outpost and seizing three prisoners, for whom U.S. forces continued to search. Respondents were informed that recent insurgent attacks had alienated the local population from the insurgents to a degree, but the population should be considered hostile to what they deem to be the U.S. occupation.

The incoming unit mission was described as neutralizing the insurgency in its area of operations. To do so, it would have to operate in dispersed, small-unit outposts to establish a security presence in population centers. At least initially, it was anticipated that operating in this pattern would invite attack, so U.S. forces must be proficient and competent from the outset. Respondents were also informed that another BCT's reconnaissance, surveillance, and target acquisition (RSTA) squadron was currently operating in the area of operations to disrupt insurgent activity and lines of communication, but would rejoin its parent unit as the incoming BCT arrives.[10]

Respondents were again informed that the scenario descriptions and characteristics of the potential incoming units were the only distinguishing elements of information available at the time of choice. They were to assume that all units were available for up to six months and were equipped at the same level, except as indicated by unit type (i.e., IBCT or ABCT).

The second-to-last data column of Table 3.2 (high-threat environment, Model 1) reports the results of the choice experiment for this scenario. As seen in the table, there are several key differences between the high-threat scenario and the other BCT exercises. First, as expected for this scenario, ABCT units were preferred on average. Second, a higher level of training for ARNG units was not a factor used to discriminate between units, indicating that respondents did not consider an additional three months of postmobilization training for ARNG BCTs to be adequate to prepare for this more-complex situation. Third, respondents—when analyzed as a group—tended to choose Regular Army BCTs—all else equal—in contrast to the moderate-threat scenario.

Consistent with the other scenarios, we tested for differences in responses between Regular Army and ARNG responses. Results, reported in the last column of Table 3.2 (Model 2), suggest the previously observed patterns hold; that is, Regular Army respondents tended to consider Regular Army BCTs to be more appropriate for this scenario, all else equal, and that reducing operational risk levels from high to low or being able to obtain a unit with three months of counterinsurgency experience could cause Regular Army respondents to choose a comparable ARNG unit. ARNG respondents apparently did not consider Regular Army units—all else equal—to be more effective in this scenario. They did tend to accord the same weight to other factors as Regular Army respondents.[11] It is interesting that ARNG respondents would prefer ARNG units with higher levels of predeployment training—suggesting that they believe that units with more training will be more effective—but do not demonstrate a similar preference for Regular Army units with even higher levels of training.

---

[10] We based our description of a high-threat scenario loosely on the situation faced by the 3rd BCT, 3rd Infantry Division in early 2007 in the "belt" area south of Baghdad, described in Dale Andrade, *Surging South of Baghdad: The 3d Infantry Division and Task Force Marne in Iraq, 2007–2008*, Washington, D.C.: U.S. Army Center of Military History, 2010. See, in particular, the description of the situation on pp. 22–24, 27, 38–41, and of insurgent activities on pp. 66–67.

[11] The test of equivalent coefficients across all covariates was rejected at a p value of 0.15, with Chi-squared test statistic of 10.70 (7 degrees of freedom), while a similar test that imposed identical coefficients across active and reserve respondents except for the *Regular Army* indicator was not rejected at a p value of 0.71, test statistic of 3.75 (6 degrees of freedom).

Qualitative Analysis

The response rate to the qualitative questions in the BCT survey was the highest across all surveys and ranged between 56 to 70 percent. The frequency distribution of answers points to variation across each experiment in respondents' perception of specific attributes and scenarios (Table 3.3). For scenario 2, specifically, comments suggesting that a specific attribute was not listed in the choice experiment comprise the most frequent category.

A closer examination of the comments in the "new attribute" category revealed that the quality of leadership constituted the most frequently mentioned characteristic that was not included in the choice experiment. The definition of the quality leadership varied across respondents. Some respondents focused on such tangible aspects of leadership as training and battlefield experience (e.g., "Experience of leadership; prior deployments—background of where and what and who they worked for"; "CIED [counter-IED] training/experience; How many battle tested leaders? How many leaders with two types of BCT experience? (IBCT and ABCT) Did they get a CTC rotation (CONUS [continental United States])?"). Others emphasized such intangible factors as morale and the ability to innovate in a complex situation (e.g., "Quality of the unit's leaders, both officers and non-commissioned officers. Observed ability of the unit [collective] and leaders to innovate in ambiguous environments based on the assessments of CTC trainers"). Less frequently mentioned attributes in this category included "Augmentation by specially trained SFATT [Security Force Assistance Transition Team] members" and "unit's morale and cohesion."

Maneuver Battalion Results

Like BCTs, maneuver battalions are responsible for conducting and integrating operations within their specified area of operations. Like BCTs, they are responsible for integrating and synchronizing effects from a diverse range of sources. Unlike BCTs, however, they have a much smaller range of organic capabilities. Most maneuver battalions

Table 3.3
Distribution of Comments Across Categories for BCT Survey

	Response Rate	Redundant	New Attribute	Scenario
Scenario 1	70% (16)	50% (8)	50% (8)	0%
Scenario 2	70% (16)	25% (4)	62% (10)	13% (2)
Scenario 3	56% (13)	54% (7)	38% (5)	8% (1)

NOTES: Raw numbers of responses in parentheses. Response rate = comments reiterating the importance of attributes already listed in the choice experiment; new attribute = comments mentioning attributes not listed in the choice experiment; scenario = comments on the clarity or plausibility of the scenario.

contain only one principal type of unit—for example, infantry companies—with other capabilities being provided by supporting units not organic to the battalion. A battalion includes about 700 soldiers.

The research team developed three scenarios for the survey of maneuver battalions. Scenarios varied with respect to operational environments, threat levels, and associated missions. To develop the low- and moderate-threat scenarios, we relied largely on interviewees' descriptions of situations in which ARNG maneuver battalions had been employed.

Unit Characteristics

As with the other surveys, we identified salient characteristics through interviews. In the maneuver battalion survey, the research team denominated opportunity cost in terms of continuity. Units available for only a brief period of time were considered to have high opportunity costs, while units available throughout the whole period of a normal deployment, nine months or longer, were considered to have low opportunity costs:

- **Component** (two levels):
  - ARNG
  - Regular Army.

- Postmobilization training (two levels, ARNG only)[12]
  - *Moderate*: Completed three months of postmobilization training sufficient for SECFOR operations conducted away from FOBs.
  - *High*: Completed nine months of post-mobilization training, including an MRE at a CTC.

- In-country experience (three levels)
  - *None*: No prior in-country experience.
  - *Three months security*: Three months of in-country experience providing installation and route security.
  - *Three months counterinsurgency*: Three months of in-country experience conducting counterinsurgency operations.

- **Continuity** (three levels):
  - *Low* (three months with no replacement for three months): Battalions would return to a theater reserve role within three months. No immediate replace-

---

[12] Postmobilization training referred to training conducted as a unit to prepare for deployment. It included training conducted while mobilized under Title 32 authority immediately prior to mobilization under Title 10, as well as training conducted after Title 10 mobilization. Regular Army units were assumed to have completed more than nine months of training as a unit, so Regular Army units' training time were indicated as "N/A."

ment would be available for them, leaving at least a three-month gap until additional U.S. forces arrive in country.

- *Moderate* (six months with immediate replacement): Battalions would be available for six months and then immediately replaced from the U.S. forces scheduled to arrive in theater at that time.
- *High* (available until relieved): Battalions would be able to remain in the area of operations until relieved by another U.S. maneuver battalion (more than nine months) or until the decision is made to transition responsibility for security to host-nation security forces.

In addition, unit type was included as a characteristic for the high-threat scenario. Combined arms battalions' greater armored protection and firepower are generally considered an advantage in such scenarios. Including this option allowed the study team to compare the value respondents placed on the tangible characteristics of superior protection and firepower to the intangible characteristics represented by units' component. This characteristic is described by:

- **Battalion type** (two levels):
  - combined arms (heavy) battalion (this is a combination of tank and mechanized infantry companies)
  - infantry battalion (a dismounted force).

Appendix A provides the survey instruments for the battalion level, including the full description of the scenarios and unit characteristics.

Respondents

Thirty-three soldiers took the Maneuver Battalion survey. As with the BCT survey, respondents brought a wealth of experience to their participation. Table 3.4 decomposes the sample by most responsible position and average months deployed.

The following subsections describe the estimation results for each of the three scenarios.

Table 3.4
Battalion Survey Sample Distribution by Most Responsible Position

Position	Regular Army		ARNG	
	Number	Average Months Deployed	Number	Average Months Deployed
Other brigade-level command	5	37.0	4	22.5
Maneuver/fires battalion commander	12	34.9	7	23.5
Deputy BCT commander	1	35.0	3	23.3

Maneuver Battalion Statistical Results

Table 3.5 shows the results for maneuver battalions for each scenario. A discussion of results for each scenario environment follows.

Low-Threat Scenario Results

In the low-threat scenario, respondents were told that they were the G-3 (operations officer) for a U.S. division conducting counterinsurgency operations in a developing nation. The operational environment was described as one in which a U.S. division, employing the theater reserve, recently crushed an insurgent uprising in one of the cities in the area of operation. The operation inflicted heavy losses on the insurgents, but insurgent networks still remained entrenched. The threat level was described as one in which insurgents were seemingly regaining strength. While their level of military capability remained low, they were exploiting poor governance and lackluster economic conditions to mobilize support from socially and economically disadvantaged elements of the local population. More established elements of society in the city, however, opposed the insurgency. Even in their weakened state, insurgent forces overmatched host-nation police and security forces. The incoming unit mission was described as one in which the incoming battalion will replace the RSTA squadron of the BCT currently responsible for this city. The BCT was employing its RSTA squadron to provide security in the city, but the unit lacked the manpower and capabilities to also support developing host-nation forces and host-nation governance. Moreover, the brigade needed the RSTA squadron to interdict enemy lines of communication into and across the province.

The first data column of Table 3.5 (Model 1) presents the results from the choice experiment for this scenario using the complete sample. All characteristics enter the model as indicator (or "dummy") variables that take on a value of one if a characteristic takes the value indicated in the "variables" column, and zero otherwise. The reference case (for which all indicator variables are assigned a value of zero)[13] is an ARNG unit with moderate levels of training, high opportunity costs in terms of continuity, and no in-country experience.

The results show that, all else equal, respondents tended to choose Regular Army units $\beta_{Active} = 1.169$. Increasing ARNG predeployment training levels was not sufficient to induce the average respondent to choose the ARNG unit $(\beta_{Active} = 1.17 > \beta_{Train-High} = 0.72)$. However, in some cases, an improvement in continuity or increase in in-country experience for an ARNG unit was generally sufficient to induce choice of an ARNG unit relative to an otherwise identical Regular Army unit. For example, respondents tended to choose an ARNG battalion with high training, low potential for continuity, and recent in-country experience conducting SECFOR operations over a Regular Army unit with low potential for continuity and

---

[13] This corresponds to a scale index value of zero.

Table 3.5
Choice Experiment Results, Maneuver Battalions

Variables	Low-Threat Environment		Moderate-Threat Environment		High-Threat Environment	
	Model 1	Model 2	Model 1	Model 2	Model 1	Model 2
Regular Army unit	1.169***	0.275	0.709***	0.0642	1.165***	0.390
	(0.305)	(0.217)	(0.267)	(0.355)	(0.266)	(0.304)
Train—high (ARNG units only)	0.716***	0.816***	0.558***	0.600***	0.446	0.471
	(0.171)	(0.174)	(0.194)	(0.194)	(0.293)	(0.298)
Continuity—moderate	2.354***	15.90***	2.778***	15.19***	0.585**	0.656**
	(0.644)	(0.395)	(0.563)	(0.420)	(0.256)	(0.267)
Continuity—high	3.195***	17.04***	3.772***	16.41***	0.990***	1.113***
	(0.608)	(0.335)	(0.587)	(0.405)	(0.307)	(0.309)
Experience—security	0.803***	1.065***	0.628*	0.744**	0.364**	0.326*
	(0.311)	(0.303)	(0.349)	(0.368)	(0.183)	(0.192)
Experience—counterinsurgency	1.365***	1.583***	1.074***	1.172***	0.754***	0.818***
	(0.289)	(0.298)	(0.362)	(0.398)	(0.231)	(0.244)
Combined Arms (Heavy)	—	—	-—	—	0.649**	0.632**
					(0.263)	(0.269)
Regular Army Respondent* Regular Army unit	—	1.724***	—	1.186**	—	1.555***
		(0.547)		(0.491)		(0.429)
Regular Army Respondent* Continuity—moderate	—	–13.52***	—	–12.54***	—	—
		(0.934)		(0.783)		
Regular Army Respondent* Continuity—high	—	–13.90***	—	–12.88***	—	—
		(0.943)		(0.696)		
Number of observations	33	33	33	33	33	33
Number of choices	495	495	495	495	495	495

NOTES: Multinomial logit model results with clustered (by individual) standard errors in parentheses. Model 1 pools all respondents. Model 2 is final restricted model (zero coefficient restrictions imposed) allowing for different coefficients by respondent component. *** $p < 0.01$, ** $p < 0.05$, * $p < 0.1$. Baseline category is ARNG battalion with moderate postmobilization training, low continuity, and no experience. Low- and moderate-threat scenario experiments did not include combined arms (heavy) attribute. Training: High variable does not apply to Regular Army units. Absolute values of coefficients are not comparable across scenarios.

no recent in-country experience, as shown by estimated scale index values of 1.5 versus 1.3, respectively.[14] In terms of relative value, all else equal, a Regular Army unit—relative to an ARNG battalion—was valued about 46 percent more than three months of in-country SECFOR experience, and just over 63 percent more than nine months of postmobilization training for an ARNG unit.[15] However, it was valued at just less than 50 percent of moderate continuity and near 37 percent of high continuity (relative to low continuity). In other words, all else equal, the average respondent tended to choose the option with the highest potential for continuity regardless of component.

Survey respondents for the battalion scenarios included both Regular Army ($n = 18$) and ARNG ($n = 15$) service members. To test whether or not preferences differed between these two groups, a model was estimated that allowed the coefficients of the choice model to vary between groups.[16] Statistical tests were performed to test if the coefficients were statistically different between groups, as described in Chapter Two. Applying the implications of these tests to the expanded model results in the estimates shown in the second data column of Table 3.5 (Model 2).[17]

There were two major results of note. First, as indicated by the nonsignificant coefficient on *Regular Army unit* (applicable to ARNG respondents) and the positive and significant coefficient on the interaction variable *Regular Army Respondent*Regular Army unit*, it was principally Regular Army respondents who considered Regular Army units to be more effective, all else equal. Nevertheless, even for Regular Army service members, the effects of increasing continuity were more important than being able to employ a Regular Army unit with no recent in-country experience or potential continuity. Similarly, Regular Army respondents considered a combination of high levels of training and counterinsurgency experience to outweigh units' Regular Army status in importance.

Second, on average, ARNG respondents were more likely to choose on the basis of continuity/opportunity costs alone (as indicated by the large coefficient values on these variables) than Regular Army respondents (see the negative and significant coef-

---

[14] The scale index value for the Regular Army unit in this case is given by the coefficient on the *Active* variable, which is 1.17. The scale index value for the ARNG unit is given by the sum of the coefficients on *Train—High* and *Experience—Security*, or $0.716 + 0.803 = 1.519$.

[15] These percentages are derived from the ratio of the relevant coefficients; that is, relative to the coefficient on counterinsurgency experience, the Regular Army coefficient is $1.169/1.365 = 0.856$, and relative to the coefficient on high levels of training is $1.169/0.716 = 1.633$.

[16] This was done by introducing interaction terms between a dummy variable that indicated a respondent was a Regular Army service member and each characteristic.

[17] Specifically, a Wald test that the six new interaction variables associated with Regular Army personnel were equal to zero was rejected at a p value of 0.0000 (Chi-squared statistic of 538.77 with 6 degrees of freedom), but a test of all but the interaction term related to component and the continuity interactions was not rejected at a significance level of 0.33 (test statistic = 3.42, 3 degrees of freedom). As such, in the table, the jointly insignificant interaction coefficients related to training, security, and counterinsurgency were set to zero.

ficients on the interaction variables at the bottom of the table). In fact, continuity was the most important contributor attribute to choices for respondents from both components.

Moderate-Threat Scenario Results

Scenario 2 was the "moderate-threat" environment. The role of the respondent was unchanged from the low-threat environment. However, the operational environment was changed to one in which several provinces in the area of operations were currently undergoing postconflict reconstruction. In one of the provinces, the insurgency had gained enough strength that operations of the resident provincial reconstruction team and agribusiness development team became high risk.

The threat level was described as one in which the insurgents mostly employed IEDs but were capable of conducting squad- and platoon-size ambushes of limited effectiveness more or less concurrently at different locations in the province. The incoming unit mission was described as one in which the incoming maneuver battalion would provide support to governance and the development of host-nation security forces. The division commander felt that these were the most important priorities. The third data column in Table 3.5 presents the results from the choice experiment for this scenario using the complete sample using the same reference case as in the low-threat scenario.

Qualitatively, the results for the moderate-threat environment were very similar to those of the low-threat environment, including, all else equal, a preference for Regular Army battalions across the sample population. As before, either three months of recent in-country experience in counterinsurgency or moving from low- to moderate- or high-continuity situations could compensate for respondents' preference for Regular Army units. Although the ratio of the coefficient on *Regular Army unit* to the other model coefficients decreased in each case, suggesting that the relative value of Regular Army units has declined with the increased threat (or alternatively, the relative value of the other components has increased), these differences were not statistically different between the two scenarios.[18]

As in the previous subsection, we tested if there is a difference in preferences between Regular Army and ARNG respondents. The final model is reported in the fourth column of Table 3.5 (Model 2). Like the low-threat environment, there was a statistical difference in coefficient estimates between Regular Army and ARNG respondents when all coefficients are tested at the 5-percent significance level.[19] This again appears to be because of the interaction between component of the respondent and the Regular Army unit type characteristic, as well as the continuity variables. Once again, either an increase in continuity from low to moderate or high, or a com-

---

[18] The p value associated with a test of equivalence was 0.73.

[19] P value of 0.0000 test statistic = 531.23, 6 degrees of freedom.

bination of high levels of training and counterinsurgency experience, was sufficient to overcome the preference for Regular Army battalions.

High-Threat Scenario Results

Scenario 3 was the "high-threat" environment. The role of the respondent was unchanged from the low-threat environment. However, the operational environment was changed to one in which an uprising has broken out in a city in a BCT's area of operations. Insurgents have seized control of the central city, including government offices and holy sites with national significance.

The threat level was described as one in which the enemy had proven capable of squad-sized maneuver and defense in urban terrain. Insurgents had employed both rocket-propelled grenades and explosively formed penetrators (EFPs) effectively. The brigade initially attempted to retake the city with an attached Stryker battalion, but the battalion took heavy losses.

The incoming unit mission was described as one in which the G-3 [the respondent] needed to reinforce the brigade with an additional maneuver battalion to enable it to resume the offensive.

The fifth data column in Table 3.5 (Model 1) presents the results from the choice experiment for this scenario using the complete sample and the same reference case as the previous two scenarios. In addition to the characteristics previously used, the alternatives included battalion type (i.e., either infantry [reference case] or combined arms [heavy]).

As in the previous two scenarios, respondents tended to choose Regular Army battalions in the high-threat environment, all else equal. In addition, the relative value of a Regular Army unit was considerably higher than in the previous scenarios, with ratios of the coefficient on *Regular Army unit* to the other coefficients considerably greater. In other words, the willingness of respondents to substitute ARNG units with high levels of predeployment training, recent in-country experience, or increased continuity for Regular Army battalions was lower in the high-threat scenario. That said, respondents tended to choose a highly trained ARNG unit with three months counterinsurgency experience and high levels of anticipated continuity over a Regular Army unit with low potential for continuity, assuming that unit type is identical. Respondents considered combined arms units—equipped with tanks and infantry fighting vehicles—to be more effective than infantry units not so equipped.

Once again, we tested whether or not Regular Army and ARNG respondents' judgment differed with respect to the relative value of Regular Army and ARNG units. Applying the implications of these tests to the expanded model results in the estimates shown in the last data column of Table 3.5 (Model 2).[20]

---

[20] A Wald test that the seven interaction variables associated with Regular Army personnel were equal to zero was rejected at a p-value of 0.0321 (Chi-squared statistic of 15.32 with 7 degrees of freedom), but a test of all but

As in the low- and moderate-threat environments, there was a significant difference in the coefficient estimates associated with *Regular Army unit* depending on the respondents' component. There is good evidence that Regular Army respondents valued Regular unit battalions as more appropriate for use in the high-threat scenario. ARNG respondents did not share this belief. There were no statistical differences in coefficients for the other characteristics in the high-threat scenario.

Unlike in the low- and moderate-threat scenarios, Regular Army respondents still valued a Regular Army battalion over an ARNG unit with high levels of predeployment training and recent in-country counterinsurgency experience. That is, Regular Army respondents were less willing to substitute training and experience for component type in high-threat scenarios. Furthermore, it was no longer the case that this preference could be overcome with an increase in continuity.[21] In short, the choice experiment revealed that Regular Army respondents did not consider ARNG maneuver battalions with increased predeployment training, recent in-country operational experience, and potential for continuity to be adequate substitutes for Regular Army maneuver battalions in high-threat environments. Precisely why the more-experienced Regular Army respondents made these valuations is not evident via the available data.

Qualitative Analysis

The response rate to the open-ended question about any additional unit attributes ranged from 45 to 72 percent. The lower rate for scenario 3 is not surprising because of possible respondent fatigue by the end of the survey. The distribution of comments across categories was similar across scenarios, with the most frequent comments relating to additional characteristics being important. Table 3.6 indicates the distribution of responses by type.

Table 3.6
Distribution of Comments Across Categories for Battalion Survey

	Response Rate	Redundant	New Attribute	Scenario
Scenario 1	72% (21)	29% (6)	66% (14)	5% (1)
Scenario 2	63% (18)	27% (5)	61% (11)	11% (2)
Scenario 3	45% (13)	38% (5)	54% (7)	8% (1)

NOTE: Raw numbers of responses in parentheses. Response rate = comments reiterating the importance of attributes already listed in the choice experiment; new attribute = comments mentioning attributes not listed in the choice experiment; scenario = comments on the clarity or plausibility of the scenario.

---

the interaction term related to component was not rejected at a significance level of 0.81 (test statistic = 2.98, 6 degrees of freedom). As such, in the table, the six jointly insignificant coefficients were set to zero.

[21] In other words, the coefficient on the *Active Respondent*Regular Army* unit variable is greater than those on *Continuity—Moderate* or *Continuity—High*.

Most comments in the "New Attribute" category focused on the quality of battalion leadership. Respondents were particularly interested in the leadership experience (e.g., "What is the level of discipline and the quality of the leadership of the various units?"; "Character of commanders"; "For those with in-country experience: assessment of performance and leadership"; "experience of command team and staff"). Other comments focused on performance and specific details about its training (e.g., "Clearer understanding of the METL [Mission Essential Task List] tasks trained after mobilization (Compo2). Track record of Compo2 units from previous deployments"); relevant civilian skills including language, knowledge of local culture and government, and development skills (e.g., "Unit must have a balance of security and reconstruction/ development skills"; "Unit Leadership, Discipline, Cohesion, Command Climate, Language/Cultural proficiency"), and intelligence capabilities (e.g., "Sensing of the units capability to gather human intelligence and its understanding of how the population is the center of gravity of the effort").

There were only four comments regarding the scenario (three of them submitted by the same respondent). One of them focused on the relevance of the choice set to the specific scenario by pointing out that combat arms basic branch units should have only been considered for scenario 1.[22] The remaining three comments touched on the design of the experiment (e.g., "The problem set seems to be really pushing for validation by a CTC exercise as a selection criteria," "This scenario moves closer to decisive action than the previous two and decreases the weight given to the duration of availability"). None of these comments posed any major concerns about respondents' ability to understand scenarios and to the internal validity of the quantitative results.

Comments in the "Redundant" category reiterated the relative importance of attributes included in the choice set (e.g., "Experience and [opportunity] cost were weighted criteria"; "Collective training and in-country experience are still factors; however, length of deployment time is now most critical for building governance and HN SF [Host Nation Security Forces]").

Findings

Stated preference analysis is based on the notion of trade-offs. The point of the analysis was not so much to determine whether respondents preferred a particular collection of characteristics (goods and services) but to use stated choices to infer how much of one characteristic they were willing to give up to get another. The more one was willing to give up, the more valuable the characteristic was to the individual. Our analysis indicated that Regular Army respondents considered Regular Army maneuver units

---

[22] In the survey, choices were either infantry or combined arms battalions. There were no choices that were not "combat arms basic branch" units.

significantly more capable than their ARNG counterparts, all else equal. Other factors could outweigh this preference once the assumption of "all else equal" was relaxed. In particular, results indicated that Regular Army respondents believed ARNG maneuver units could be employed in low- to moderate-threat level environments at acceptable levels of risk, if suitable risk mitigation measures have been undertaken (i.e., more predeployment training and time to gain in-country experience).

Regular Army and ARNG Officers Differed Starkly About the Relative Effectiveness of Regular Army and ARNG Maneuver Units

Our statistical analysis confirmed what we suspected before conducting the survey. Regular Army respondents valued Regular Army units as significantly more capable than their ARNG components, all else equal, while ARNG respondents assigned no significant difference in capability between the two.

Regular Army respondents accorded greater weight to units' Regular Army status than to almost any other factor except for operational risk, especially in determining which maneuver battalions or BCTs to employ. For maneuver battalions, the component factor weighed more heavily than the nature or degree of units' prior experience in theater, units' amount of predeployment training time or even the unit type—combined arms or infantry—in a high-threat environment. All other things being equal, Regular Army respondents valued Regular Army maneuver units as the best choice at the battalion and BCT level; we did not conduct a similar analysis at company or platoon levels.

Operational Risk Was the Most Important Factor in Selecting BCTs

In the choice experiment for BCTs, we used operational risk as a proxy for opportunity costs of selecting one unit over another. We denominated those opportunity costs in terms of operational risk, defined as the chance that the campaign might suffer moderate to severe setbacks if resources were diverted to the scenario at hand from more important missions. As indicated by the weights associated with each factor, respondents considered operational risk to be the most important factor in selecting BCTs. In low- and moderate-threat scenarios, Regular Army respondents were willing to employ ARNG units rather than disrupt the overall campaign plan significantly by diverting a Regular Army BCT from a higher-priority mission. In high-threat scenarios, however, Regular Army respondents were willing to accept disruptions to the overall plan in order to be able to employ a Regular Army BCT. In the high-threat scenario, it can be inferred that Regular Army officers believed they would incur at least as much risk by employing an ARNG BCT as they would by diverting a Regular Army unit from another area of operations.

Continuity Was the Generally Most Important Factor in Selecting Maneuver Battalions

For the maneuver battalion choice experiment, we defined opportunity cost in terms of continuity. Respondents preferred options with longer periods of continuity. All respondents preferred units with more time remaining in country to those with less time. Units that could remain in the area of operations for nine months or longer were preferred to units that would be replaced after six months. Almost any option was preferred to one in which a unit would be withdrawn without replacement for three months. Respondents preferred units with a high degree of continuity to any other factor—including predeployment training, component status, and recent in-country experience in low- and moderate-risk scenarios. In high-threat environments, however, it was more important to Regular Army officers to employ a Regular Army unit than it was to maximize continuity. In short, it appears that Regular Army officers assessed greater risk from employing ARNG units in a high-threat scenario than they did from accepting a gap in coverage.

Other Factors Could Outweigh Regular Army Status, Especially in Combination

As we have indicated, the operational risk associated with a particular option outweighed Regular Army status in the unit scenarios, even for Regular Army respondents. Combinations of those factors could offset the preference for Regular Army status with respect to a particular choice. Even in a high-threat scenario, in which significant combat was anticipated, Regular Army respondents would have chosen an ARNG combined arms battalion with in-country experience conducting counterinsurgency operations elsewhere and six months or more of predeployment training over a Regular Army infantry battalion just arriving in theater. Similarly, Regular Army respondents would choose an ARNG ABCT with previous in-country experience over a Regular Army IBCT just arriving in theater in the high-threat scenario. Even for Regular Army respondents, there were more important considerations in selecting units for employment than a unit's component of origin.

It would be difficult to replicate some of these factors in the real world. For example, it would probably be impractical for incoming units to cycle through low-risk areas of operation upon arrival in theater to prepare for their more demanding, follow-on assignments. In real terms, such a practice would impede units' continuity and understanding of their areas of operation, both being critical factors for counterinsurgency operations. Even in theoretical terms, advantages from such a practice would also accrue to newly arrived Regular Army and ARNG units, likely reinforcing rather than mitigating the perceived Regular Army advantage.

Instead, we can take this to indicate that higher levels of recent in-country experience in counterinsurgency operations—however obtained—might be able to compensate for ARNG component status. It is significant, however, that even six months of predeployment training—a figure that includes pre- and postmobilization time—did not cause Regular Army respondents to value Regular Army and ARNG maneuver units equally in terms of operational performance and risk. That period exceeds the 90 days that ARNG leaders had asserted was adequate to prepare ARNG BCTs for operations, regardless of complexity or threat. It was approximately equal to the time ARNG maneuver brigades took to train for counterinsurgency missions in 2004–2006. Our analysis thus suggests that such units would require even more pre-deployment training time to compensate for ARNG component status, although it did not allow us to estimate how much more time was required. Future research could vary training levels across different contexts to obtain an estimate of how much training time would be perceived as required to overcome other deficiencies.

Unit Key Leaders' Experience and Skills Are Critical

Regular Army officers' judgment that Regular Army units were usually the better choice seemed to stem from their assessment of the prevailing level of operational experience and skill among key leaders in Regular Army and ARNG units. Leadership emerged as a key issue from the interviews and focus groups the study team conducted to develop and refine our survey instruments. In those contexts, respondents were not using the term "leadership" to denote issues of character, charisma, or general management acumen. Rather, they used the term to indicate technical and tactical skill in the synchronization of diverse military capabilities in rapidly evolving operational environments. Around 40 percent of all comments we received from survey respondents testified to the importance of such leadership issues, especially in the core command team of commander, command sergeant major, and operations officer, among others.

In general, the Regular Army officers with whom we spoke believed that ARNG leaders had accumulated considerably less meaningful experience in the synchronization and integration of combat operations in either training or overseas operations than their Regular Army counterparts and consequently would be less capable of doing so in ongoing contingency operations. Their value of the impact of experience on proficiency was consistent with the theory on the development of expertise, in which relevant experience plays the dominant role.[23] Ergo, reducing the perceived gap between Regular Army and ARNG units' capabilities would require significantly increasing ARNG key unit leaders' level of relevant operational experience. Doing so might pose

---

[23] See, for example, John Bransford, *How People Learn: Brain, Mind, Experience and Schools*, Washington, D.C.: National Academy Press, 1998; K. Anders Ericsson, "The Influence of Experience and Deliberate Practice on the Development of Superior Expert Performance," in K. Anders Ericsson, Neil Charness, Paul J. Feltovich, and Robert R. Hoffman, eds., *The Cambridge Handbook of Expertise and Expert Performance*, Cambridge, UK: Cambridge University Press, 2006; K. Anders Ericsson, Ralf Th. Krampe, and Clemens Tesch-Römer, "The Role of Deliberate Practice in the Acquisition of Expert Performance," *Psychological Review*, Vol. 100, No. 3, 1993, pp. 363–406; Gary Klein, *Sources of Power: How People Make Decisions*, Cambridge, Mass.: Massachusetts Institute of Technology Press, 1999; Robert G. Lord and Karen J. Maher, "Cognitive Theory in Industrial and Organizational Psychology," in Marvin D. Dunnette and Leaetta M. Hough, eds., *Handbook of Industrial and Organizational Psychology*, Vol. 2, Palo Alto, Calif.: Consulting Psychologists Press, 1991, pp. 1–62.

a challenge to part-time ARNG leaders who also have full-time civilian jobs or perhaps require a different manning model that assigns full-time personnel (Regular Army and/or full-time support) to key ARNG leader billets.

It is also consistent with a conceptual model developed in a previous RAND Arroyo Center study on National Guard Special Forces, as shown in Figure 3.1. In this model, the red starbursts symbolize deployments, the green curve represents a Regular Army officer's career trajectory, while the blue curve represents an ARNG officer's career trajectory. In fact, one could substitute major training events or any other major operational experience for the deployments. The model of Figure 3.1 posits that officers' proficiency increases with major developmental experiences over time. Over the same period of time, Regular Army officers will accumulate more deployments and relevant operational experience. Presumably, as experience increases, so will expertise.[24] Note that the model shown in Figure 3.1 is entirely theoretical. A measure of aggregate proficiency has yet to be developed, and the relationship between experience and proficiency has yet to be quantified except in a more or less intuitive "more is better" sense.

That, at least, is the theory, which is consistent with theory on the development of expertise. We lack objective data on performance, however, and even data on the posi-

Figure 3.1
The Role of Deployments in Building Leadership Proficiency

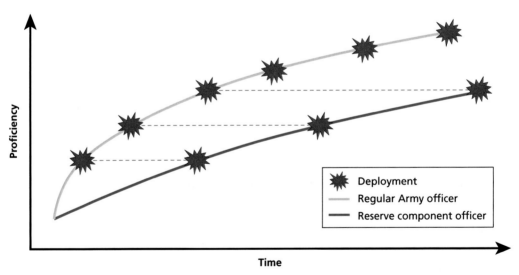

SOURCE: Peters, Shannon, and Boyer, 2012, Fig. 2.12.
RAND RR1745-3.1

---

[24] John E. Peters, Brian Shannon, and Matthew E. Boyer, *National Guard Special Forces: Enhancing the Contributions of Reserve Component Army Special Operations Forces*, Santa Monica, Calif.: RAND Corporation, TR-1199-A, 2012, p. 19.

tions soldiers occupied when deployed is fairly difficult to synthesize. Ergo, that theory is yet to be tested empirically.

We did not include leadership as a variable in most of our maneuver unit scenarios, however. Instead, respondents presumably included their assumptions about the prevailing levels of leadership proficiency in the Army's three components in weighing the importance of component in the decision. If this was indeed the case, it might help explain Regular Army officers' preference for Regular Army combat units, especially for high-threat scenarios.

Regardless of whether ARNG respondents would agree about the degree to which experience affects proficiency, they certainly seemed to agree that the skills of key leaders had a disproportionate impact on units' operational effectiveness. It is thus reasonable to deduce that improving key leaders' proficiency should be a focus for all Army units.

Conclusion

This chapter explained the analysis of stated preference survey data with respect to maneuver units in counterinsurgency. That analysis indicated that Regular Army and ARNG respondents differ significantly in their respective assessments of Regular Army and ARNG maneuver units. Regular Army respondents valued Regular Army units as significantly more capable, while ARNG respondents reported no differences. Still, other factors can combine to outweigh Regular Army officers' preferences. For all respondents, however, the single most important unit characteristic is opportunity cost, whether in terms of the risk a choice poses for the overall operational plan or in terms of continuity associated with their choice. All respondents identified leadership as an important factor in determining units' combat effectiveness. In the next chapter, we will perform similar analyses with respect to enablers.

CHAPTER FOUR

Enablers

This chapter assesses the degree to which the Army can rely on selected ARNG and USAR enablers in counterinsurgency operations. By way of reminder, an *enabler* is a unit that supports maneuver units in the conduct of operations. Like the previous chapter, this one describes the statistical analysis, explains the results, and synthesizes findings.

Regular Army respondents' preference for Regular Army enabler units was considerably more muted with respect to engineer and MP units. In fact, Regular Army respondents only indicated a preference for Regular Army engineer companies, a preference that could be overcome by combinations of other unit attributes. As with maneuver units, operational risk and leadership were also important attributes.

Engineer Survey Results

Survey respondents who were qualified to answer questions about engineer battalions and companies were asked to complete one choice experiment for each unit type. The choice experiment for engineer battalions assessed those units' ability to provide C2 for engineer operations. The experiment about companies assessed the ability to actually conduct a certain set of operations. We start with the battalion survey. Keep in mind that, under the Army's modular organization, an engineer "battalion" is actually just the headquarters that provides C2 for a number of different engineer assets, the type and number of which will vary with the situation.

Engineer Battalion Options' Characteristics

Under the Army's modular organization, an engineer battalion is a C2 headquarters with no organic engineer companies. Doctrinally, engineer battalions can control up to five engineer companies of different types. Those different types include sappers, route clearance companies, and horizontal and vertical construction companies.[1] An

---

[1] For a description of the functions and organization of the engineer battalion, see Headquarters, Department of the Army, "Engineer Operations," Washington, D.C., FM 3-34, August 2011, pp. 2-8 through 2-9.

engineer battalion is found at echelons above brigade; it is not to be confused with a *brigade engineer battalion*, an organization that recently replaced the special troop battalions in BCTs. The term *battalion* may seem counterintuitive for those used to thinking of battalions as consisting of several hundred soldiers organized to perform a single function. For the purposes of this analysis, we focused on the C2 of engineer operations, including route-clearance and construction operations. Route clearance is a mission in which units are to find and neutralize hazards such as obstacles, mines, and IEDs. Construction operations involve building roads and buildings, just as it sounds.

The characteristics and potential levels for the engineering battalion experiment are below. For this survey, we reverted to using operational risk as the opportunity cost variable. The engineer battalion survey was the one in which we made leadership, or C2 capability, an explicit variable.

- **C2 capability** (three levels): Battalion commanders and staff vary in their capability to C2 engineer operations. Respondents were told that values represented a notional assessment made by them in consultation with trusted subordinates (e.g., command sergeant major) of the battalion headquarters' capability to C2 operations. Respondents formed these assessments based on site visits to these battalion headquarters' training sites as they prepared for deployment.
  - *Low confidence:* Doubt the ability of the battalion commander and staff to C2-demanding combat operations.
  - *Moderate confidence:* The battalion commander and staff were adequate to perform their anticipated mission. There was uncertainty about their ability to cope with unexpected increases in the threat level or changes to enemy tactics.
  - *High confidence:* The battalion commander and staff were capable of providing C2 to subordinate units' engineer operations under any conceivable circumstances in the area of operations.

- **Operational risk** (three levels): A battalion headquarters selected for the respondent's mission cannot be employed anywhere else in theater. This creates costs to the allied mission in terms of potentially increased risk.
  - *None: just deployed.* This was an incoming battalion headquarters scheduled to replace the outgoing battalion.
  - *Moderate: Battalion headquarters is being remissioned from a quieter sector.* The battalion headquarters was currently supporting two brigades in a quieter sector controlled by the allied forces and would be reassigned from there to the respondent's mission. While diverting this battalion headquarters could increase friction, it would not significantly increase risk in the allied unit's area of operations.
  - *High: Battalion headquarters is being remissioned from an active sector.* The battalion headquarters was currently supporting three brigades in a strate-

gically important sector in an active area of operations and would be reassigned from there to the respondent's mission. Diverting this battalion headquarters would incur a significant increase in risk in that area of operations.

- **Effectiveness** (two levels): Battalions would have different levels of observable in-country experience.
  - *Satisfactory performance in postmobilization training, no recent in-country experience.* The battalion headquarters' performance in postmobilization route clearance training was satisfactory, but there is no additional information.
  - *Demonstrated effectiveness in theater.* Indicates good performance in their current role.

- Force structure component (two levels):
  - U.S. Army Reserve
  - Regular Army.

Appendix A provides the survey instruments for the engineer battalion experiment, including the full description of the scenario and unit characteristics. Eleven individuals answered each of the engineering battalion questions. Because of this small sample size, results should be interpreted with caution.

Respondents
Of the 11 engineer officers who responded to this survey, four were Regular Army and seven were ARNG and USAR officers. The small number of respondents should inspire caution in generalizing from these results, although the standard errors of the coefficient estimates take this small sample size into account. Table 4.1 breaks out respondents by most responsible position held when deployed and average number of months deployed.

Battalion Scenario Results
Respondents to the engineer battalion scenario were told that they were the commander of an engineer brigade supporting a multinational coalition conducting counterinsurgency operations in a developing country. The task was to choose the most appropriate battalion headquarters for the operational environment and threat level provided in the scenario. Respondents were asked to assume that all battalion headquarters were available at the time needed and appropriately equipped for the mission.

The operational environment was described as one in which enemy activity had declined significantly in this region following aggressive clearing operations by the supported BCTs. The supported BCTs would be consolidating their gains, and demand

Table 4.1
Engineer Survey Sample Distribution by Most Responsible Position

	Active		Guard		Reserve	
	Count	Average Months Deployed	Count of Grade	Average Months Deployed	Count of Grade	Average Months Deployed
Maneuver/fires battalion commander			2	16		
Other battalion-(presumably engineer) level command	3	24	2	16	1	13
Other brigade-level command	1	35	1	26	1	12
Total	4	27	5	18	2	13

for construction support would probably increase, especially in support of host-nation security forces.

The threat level was described as one in which direct-fire enemy contact is infrequent. The enemy prefers to engage using IEDs, of which about half were command-wire detonated.

The incoming battalion headquarters mission was to replace one engineer battalion headquarters that was due to rotate back to the United States in two months. The incoming battalion headquarters would provide C2 over three route-clearance companies, two of which were from the Regular Army, and three construction companies, all of which were ARNG or USAR.

Table 4.2 shows the results from the choice experiment for the engineer battalion choice experiment using the complete sample. All characteristics enter the model as indicator (or "dummy") variables that take on a value of one if a characteristic takes the value indicated in the "Variables" column, and zero otherwise. The reference case (for which all indicator variables are assigned a value of zero) is a USAR engineer unit with low C2 capabilities, high levels of operational risk, and satisfactory performance in postmobilization training, but no experience in theater.

The results show that neither effectiveness nor component could be identified as characteristics on which respondents discriminated in the choice process, but that C2 capabilities and associated levels of operational risk were significant determinants that entered with the expected sign. While lower levels of risk are preferred, the precision of the estimates is insufficient to conclude that there were differences between moderate and low levels of risk at a 95-percent level of confidence.[2] Nevertheless, it appears that respondents tended to pick the unit with a greater C2 capability unless associated levels of operational risk are lower than the baseline.

---

[2] Formal test of equivalence has p value of 0.0576.

Table 4.2
Choice Experiment Results, Engineering Battalion

Variables	Model 1
C2 capability—moderate	2.293*
	(1.330)
C2 capability—high	3.650**
	(1.614)
Associated level of operational risk—moderate	3.479***
	(1.219)
Associated level of operational risk—low	2.757**
	(1.174)
Effectiveness—demonstrated	−0.0588
	(0.409)
Regular Army unit	0.0294
	(0.478)
Number of observations	11
Number of choices	165

NOTES: Multinomial logit model results for engineering battalion experiment with clustered (by individual) standard errors in parentheses. *** $p < 0.01$, ** $p < 0.05$, * $p < 0.1$. Baseline category is USAR engineering unit with low C2 capabilities, high associated levels of operational risk, and satisfactory performance in postmobilization training, but no experience in theater.

Given the low sample size, we were unable to test differences between Regular Army ($n = 4$) and ARNG or USAR ($n = 7$, five ARNG and two USAR) respondent preferences, although a model that allowed for differing coefficients on component between the two groups did not detect significant differences in preferences for this characteristic.[3]

Engineer Company Characteristics

For this analysis, we focused on the route clearance company. The company's mission was to "detect and neutralize explosive hazards along routes and within designated areas as part of a tailored force package or task organized to the BCT."[4] They comprised fewer than 200 soldiers.

---

[3] *P*-value for test of equivalence was 0.298.

[4] Headquarters, Department of the Army, 2011, p. B-8.

We used a different set of unit characteristics to describe options in the company scenario. Again, we described opportunity cost in terms of the operational risk of diverting another unit from its assigned mission.

The characteristics and potential levels for the engineering company experiment included:

- **Operational risk** (three levels): A company selected for the mission could not be employed anywhere else in theater. This creates costs to the coalition mission in terms of potentially increased risk.
  - *None: just deployed.* This was an incoming company scheduled to replace the outgoing company.
  - *Moderate: remissioned from quieter sector.* The company in question would be remissioned from a quieter sector, in which it was supporting an allied brigade. While diverting this company would increase friction, it would not significantly increase risk in the allied unit's area of operations.
  - *High: remissioned from active sector:* This company was currently supporting another BCT in a strategically important area of operations in which enemy threat levels are fairly high. Diverting this company would incur a significant increase in risk in that area of operations.

- **Manning level** (three levels): Manning level was based on requirements for conducting route clearance in areas with command wire–detonated IEDs; higher levels of manning increase the company's ability to conduct dismounted operations to locate and disrupt the individual or unit that is overwatching the IED.
  - *Low, 26–30 members:* Platoon average is 26–30 members.
  - *Acceptable, 31–35 members:* Platoon average is 31–35 members.
  - *Optimal, 36–40 members:* Platoon average is 36–40 members.

- **Certification** (two levels): Reflects the ratio of assigned personnel qualified to operate key mine clearance systems to the number of personnel required to operate them on any given day.
  - *Below average:* Indicates less than 150 percent.
  - *Above average:* Indicates more than 250 percent.

- **In-country experience** (three levels): Indicates the amount of recent in-country experience possessed by each option.
  - *Low:* Indicates that a company had not completed validation and has been in the country for two weeks or less.
  - *Medium:* Indicates that a company had completed validation and has been in the country for about one month.
  - *High:* Indicates that a company had completed validation and has been in the country for three months.

- Force structure component (two levels):
    - USAR
    - Regular Army.

Appendix A provides the survey instruments for the engineer company experiment, including the full description of the scenario and unit characteristics. Eleven qualified individuals answered each of the engineer company questions. Because of this small sample size, results should be interpreted with caution.

Company Scenario Results

Respondents to the engineer company scenario were told that they were the operations officer for an engineer brigade supporting a coalition engaged in ongoing counterinsurgency operations in a developing country. The task was to choose the most appropriate engineer company for the operational environment and threat level provided in the scenario. Respondents were asked to assume that all companies were available at the time needed and appropriately equipped for the mission.

The operational environment was described as one in which the insurgent presence was more pervasive than originally anticipated. The enemy's primary mode of engagement was employing IEDs. The insurgents were continuously adapting both their IEDs and their methods of employment, whereas coalition forces continuously introduced new and improved mine detection and clearance systems and constantly adapted their tactics as well, which required keeping engineer qualifications up to date.

The threat level was described as one in which the density of IEDs in this area of operations was relatively high; most were command-wire detonated. Respondents were informed that the enemy sometimes overwatched IED emplacements with direct fire weapons.

The incoming company mission was to replace one of the subordinate route-clearance companies that was scheduled to redeploy within the next two months. The incoming company would support a U.S. Regular Army BCT that had initiated aggressive operations to clear a pervasive insurgent presence from its area of operations.

Model 1 of Table 4.3 shows the results from the choice experiment for the engineer company choice experiment using the complete sample. All characteristics enter the model as indicator (or "dummy") variables that take on a value of one if a characteristic takes the value indicated in the "Variables" column, and zero otherwise. The reference case (for which all indicator variables are assigned a value of zero) is a USAR engineer unit with high associated levels of operational risk, low manning levels, below-average certification, and low levels of recent in-country experience.

The results showed a very imprecisely estimated preference function, because of the large number of characteristics and levels compared with the sample size.[5] There

---

[5] Achievable sample samples were unknown at the time of survey design.

Table 4.3
Choice Experiment Results, Engineer Company

Variables	Model 1	Model 2
Associated level of operational risk—moderate	0.106	0.0864
	(0.461)	(0.437)
Associated level of operational risk—low	1.205	1.368*
	(0.825)	(0.780)
Manning—acceptable	1.172*	1.306**
	(0.599)	(0.571)
Manning—optimal	0.825**	0.838**
	(0.378)	(0.411)
Certification—above average	2.019***	2.208***
	(0.736)	(0.705)
Experience—medium	0.6	0.710*
	(0.386)	(0.431)
Experience—high	2.716***	3.034***
	(0.969)	(1.024)
Regular Army unit	−0.29	−1.009**
	(0.585)	(0.469)
Regular Army respondent *Regular Army unit	—	1.836**
		(0.932)
Number of observations	11	11
Number of choices	165	165

NOTES: Multinomial logit model results for engineering company experiment with clustered (by individual) standard errors in parentheses. *** $p < 0.01$, ** $p < 0.05$, * $p < 0.1$. Baseline category is USAR engineering unit with high levels of associated operational risk, low manning levels, below average certification, and low levels of in-country experience.

is some evidence that respondents discriminated on operational risk, manning levels, certification, and experience, although statistical power is low.

The second data column in Table 4.3 (Model 2) shows the same model allowing for the coefficients on *Regular Army Unit* to vary across active ($n = 4$) and ARNG and USAR ($n = 7$, five ARNG and two USAR) respondents. It does appear that preferences differed across respondent type, as given by the positive and statistically significant coefficient on the interaction term. Additionally, there is some evidence that ARNG and USAR respondents had a preference for USAR units, exacerbating the differences.

Qualitative Analysis

The number of respondents who took the engineer company survey was the lowest of all the surveys offered, and subsequently there were only three responses to the open-ended question after each scenario (Table 4.4). For scenario 1, these comments focused on unit leadership, cohesion, the relevance of predeployment training to the mission, and the number of past deployments. For scenario 2, the comments focused on leadership, supply accountability, time in country, and Medical Protection System qualification. There was also one comment questioning whether one of the choice occasions for scenario 2 was realistic because a unit had a low experience and at the same time a high level of operational risk (e.g., "The company scenarios have several conflicts—units can't have low experience and be remissioned from an active sector. Please relook this.").

Military Police Survey Results

MP companies have a number of potential missions, only one of which is law enforcement. In combat operations, MP companies also can provide area and route security.[6] Our analysis focused on the route security mission.

One scenario was developed at the MP company level. Our analysis of the deployment data indicated that few MP units had been employed at the battalion and especially brigade level, implicitly limiting the number of respondents who would have been able to assess Army units' capabilities based on their personal experience or observation.

Table 4.4
Distribution of Comments Across Categories for Engineers Survey

	Response Rate	Redundant	New Attribute	Scenario
Scenario 1	62% (5)	20% (1)	80% (4)	0%
Scenario 2	3%	33% (1)	33% (1)	33% (1)

NOTES: Raw numbers of responses in parentheses. Response rate = comments reiterating the importance of attributes already listed in the choice experiment; new attribute = comments mentioning attributes not listed in the choice experiment; scenario = comments on the clarity or plausibility of the scenario.

---

[6] Headquarters, Department of the Army, "Military Police Operations," Washington, D.C., FM 3-39, 2010, p. 2-10.

MP Company Characteristics

MP company characteristics and potential levels for the experiment included:

- **Risk to overall mission** (three levels): The additional risk that selecting a particular option would incur in terms of the commander's overall plan, particularly in terms of impact on the forthcoming offensive.
  - *No additional risk:* The company was deploying to replace the outgoing company.
  - *Moderate risk:* The company described was currently supporting an allied brigade in a quieter sector. Remissioning the company would somewhat increase friction and increase risk, but it would unlikely fundamentally alter the situation in the allied brigade's area of operations.
  - *High risk:* The company was providing effective support to a U.S. brigade in a strategically important area of operations. Diverting the company could significantly disrupt progress, especially in developing host-nation police forces.

- **In-country experience** (three levels): All units were available for the mission for up to six months, but they had varying degrees of in-country experience.
  - *None:* The company was scheduled to arrive next week.
  - *Three months:* The company had been in country for three months.
  - *Six months:* The company had been in country for six months.

- Force structure component (two levels):
  - USAR
  - Regular Army.

- **Personnel level** (two levels): Every assigned MP company must have the ability to organize and field sufficiently manned SFATTs at the squad level to meet all host-nation training requirements. Manning levels were labeled as "undermanned" or "fully manned."[7]
  - *Undermanned:* 85 percent of authorization.
  - *Fully manned*: 100 percent of authorization.

- **Validation** (two levels): Given the complexity and priority of the SFATT mission, the MP brigade commander had instituted a company-level training validation program to ensure all MP companies understand the appropriate training techniques for use with their host-nation national police/host-nation border police counterparts.

---

[7]   In the description of the unit attributes shown to respondents, it was stated that manning levels would be labeled as "undermanned," "acceptable," or "fully manned." However, only "undermanned" and "fully manned" were used and fully described in the actual experiment.

- *Yes, completed.* Indicates that a company completed the validation program.
- *No, not completed.* Indicates that a company did *not* complete the validation program.

Appendix A provides the survey instruments for the MP level, including the full description of the scenarios and unit characteristics.

Thirty-four qualified individuals answered each of the MP survey questions. They included individuals in the MP branch and battalion commanders and above from logistics branches. We had respondents in the latter category take this survey because they would have had extensive experience with MPs providing area and route security. Table 4.5 shows the sample by position and average time deployed.

MP Company Scenario

Respondents to the MP scenario were told that they were the commander of an MP battalion supporting several BCTs conducting counterinsurgency operations in a medium-sized developing nation. The task was to choose a company to replace a company scheduled to deploy in four weeks that was assigned to the battalion. Respondents were asked to assume that all potential replacement units were available for the mission for up to six months.

The operational environment was described as one in which the MP battalion was supporting ongoing counterinsurgency operations. The battalion was also providing oversight of a SFATT to a host-nation national police brigade to increase its capacity to conduct sustained independent operations and to secure the local population.

The threat level was described as one in which the density of IEDs in the area of operations was relatively high and represented the most dangerous threat to both U.S. and host-nation forces.

Table 4.5
Military Police Survey Sample Distribution by Most Responsible Position

Position	Active		Guard		Reserve	
	Count	Average Months Deployed	Count	Average Months Deployed	Count	Average. Months Deployed
Other brigade-level command					3	29.3
Maneuver/fires battalion commander	1	33.0	1	44.0		
Other battalion-level command	9	34.2	2	7.0	1	33.0
Division staff officer	6	24.0	2	22.5	7	27.1
Force management staff officer					2	8.0
Total	16	30.3	5	20.6	13	25.2

The primary task entailed the organizing and fielding of squad-level SFATT focused on the mentoring of local host-nation units, IED identification, patrolling techniques, and community policing techniques (i.e., enforcement of local laws). Respondents were informed that one of the MP companies engaged in supporting an engineer company engaged in combined-arms route clearance for the area of operations was scheduled to be relieved from theater the next month, and the task was to recommend a replacement. The mission of the incoming MP company would be focused on supporting route clearance and training host-nation counterparts on these techniques.

Respondents were informed that the scenario descriptions and characteristics of the potential incoming units were the only distinguishing information available at the time of choice. They were to assume that all units are available for up to six months and were armed with the equipment required for the mission.

Table 4.6 shows the results from the choice experiment for the MP choice experiment using the complete sample. All characteristics entered the model as indicator (or "dummy") variables, which take on a value of one if a characteristic takes the value indicated in the "Variables" column and zero otherwise. The reference case (for which all indicator variables are assigned a value of zero) was an undermanned USAR MP unit with high levels of associated operational risk that has not completed the validation training.

The results show that all unit characteristics, except component, were statistically significant and of the expected sign. As such, respondents generally did not use component to discriminate between choices. In addition, completion of validation training was viewed essentially as a substitute for being fully manned,[8] and reducing levels of associated operational risk from high to moderate was more valuable than reducing

Table 4.6
Choice Experiment Results, MP Company

Variables	Model 1
Associated level of operational risk (moderate)	3.875***
Associated level of operational risk (low)	4.392***
Regular Army unit	0.286
Personnel level (full)	1.148***
Validation training (complete)	1.488***
Number of observations	34
Number of choices	510

NOTES: Multinomial logit model results for MP experiment with clustered (by individual) standard errors in parentheses. *** $p < 0.01$, ** $p < 0.05$, * $p < 0.1$. Baseline category is USAR company with high-cost, undermanned personnel level, and incomplete validation training.

---

[8]   A test of equivalence between these two coefficients cannot be rejected at a p value of 0.16.

risk from moderate to low. However, operational risk was clearly viewed as the most valuable attribute among respondents. Furthermore, there is no evidence that Army respondents from different components (5 Regular Army, 5 ARNG, and 13 USAR) had differing preferences.[9]

Qualitative Analysis

The response rate to the open-ended question for MP was 84 percent, with 70 percent of pertaining to unit attributes. Of this category, about two-thirds focused on experience of the unit leadership. Some respondents wanted to know whether the unit leadership had combat experience leading a specific unit (e.g., "Prior deployment experience at the platoon and company levels for senior leaders; prior related training experiences of senior leaders at both PLT [platoon] and CO [company] levels"; Commander and 1SG [first sergeant] experience, temperament, etc."; "prior deployment experiences of senior NCOs [noncommissioned officers] and officers"). Others focused on the units that were undermanned and wanted to know whether the key leadership positions were staffed for them (e.g., "Key leadership positions filled with authorized rank/ grade?" "For units that are undermanned, are key leadership positions filled?"). Other comments about the leadership focused on such intangible qualities as competence and maturity (e.g., "Competence of company leadership"; "Performance of the Company leadership team during their current deployment, as applicable").

The remaining comments in the "New Attribute" category focused on previous deployment (e.g., "Prior combat experience Time unit has been training together prior to deployment"); non-combat skills including language and civilian skills and equipment (e.g., "What civilian skills do the unit members have?" "Does units [sic] have language capabilities?") and intangible measures, including morale and maturity (e.g., "level of discipline and morale in the prospective companies," "Stability of leadership, experience and maturity of personnel, equipment readiness rates, morale of personnel, etc."). Table 4.7 shows the distribution of such responses.

Three comments regarding the scenario focused on the plausibility of the description of the operational risk associated with the options. Two respondents questioned the need to divert the company from the ongoing operation (e.g., "There is almost never a reason in my experience to disrupt one sector for another if there is a *qualified* unit available to fill a position"; "[Opportunity] Cost row was confusing: why would we bring 2 companies over to replace 1?"), and the other wanted to know more about the political implications of deploying a unit.

---

[9] Test of equivalent coefficients cannot be rejected with p value of 0.52, with likelihood ratio statistic of 4.19 (5 degrees of freedom). A model that imposes identical coefficients except for *Active* results in an insignificant coefficient on *Active Respondent*Active* (p value of 0.163), implying no significant differences between active and reserve respondents.

Table 4.7
The Distribution of Comments Across Categories for MP Survey

	Response Rate	Redundant	New Attribute	Scenario
Scenario 1	84% (27)	11% (5)	70% (19)	19% (3)

NOTES: Raw numbers of responses in parentheses. Response rate = comments reiterating the importance of attributes already listed in the choice experiment; new attribute = comments mentioning attributes not listed in the choice experiment; scenario = comments on the clarity or plausibility of the scenario.

Findings

Respondents Indicated No Significant Preference with Regard to Component for Military Police Companies or Engineer Battalion Headquarters

In contrast to the other unit surveys, respondents to the MP unit surveys indicated no strong preference for units from one component or another. Many of those surveyed were not, in fact, members of that branch and may not have been sensitive to the differences between similar units from different components. Finally, we may not have been able to identify and describe the determinants of operational effectiveness to a degree that rendered component status of minimal importance.

As for engineer battalions, analysis of the 11 responses received revealed that Regular Army respondents indicated no statistically significant preference for Regular Army engineer battalions to provide C2 for engineer operations. Given the small sample size, the ability to detect such differences statistically is low.

Regular Army Respondents Preferred Regular Army Engineer Companies

In contrast, Regular Army respondents did indicate a preference for Regular Army engineer companies. The sample size was extremely small, however. Moreover, the preference—although statistically significant—was not as strong as it was for demonstrated proficiency during a certification exercise or previous experience in country.

Operational Risk Was Usually—but Not Always—the Dominant Factor

To a lesser degree than indicated by analysis of the maneuver unit surveys, operational risk was an important factor in selecting units for employment. For engineer battalions, operational risk was slightly more important than C2 capability. For MP companies, operational risk was about three times more important than any other factor. For engineer companies, however, cost was less important than previous operational experience and soldiers' certifications on key engineer systems and was of equivalent importance to maintaining adequate manning levels.

Key Leaders' Experience and Skills Are Critical

The engineer battalion survey was the only survey that explicitly included key unit leaders' capability to C2 military operations as a variable. Although the sample was small, it is notable that respondents effectively treated that leadership factor as approximately equal to associated levels of operational risk in importance. Once again, a significant number of the qualitative responses to both surveys explicitly mentioned some aspect of leadership skills as important.

Conclusion

According to this analysis, respondents expressed considerably less preference for Regular Army enablers in support of combat forces than was observed with regard to Regular Army maneuver units. To some degree, this may be a function of relatively small samples.[10] Under the right combination of circumstances, even Regular Army respondents generally found it acceptable to employ USAR engineer and MP units in the counterinsurgency scenarios described in the survey, as long as the option possessed such key characteristics as acceptable levels of manning or recent in-country experience. It would be a mistake, however, to generalize findings to other types of enablers not analyzed. Aviation units, in particular, are considerably more complex, making it much more difficult to upgrade unit capabilities in preparation for deployment. The following chapter explains our assessment of the degree to which the Army could rely on ARNG or USAR soldiers to fill key billets in strategic-level headquarters.

---

[10] A smaller sample implies less information in the data and thus noisier coefficient estimates. This impairs the ability to find differences in coefficient estimates.

CHAPTER FIVE

Individuals

This chapter explains our assessment of the degree that the Army can rely on individual Guard and Reserve officers to fill billets in contingency operations headquarters. The U.S. Department of Defense (DoD) does not organize such headquarters in peacetime and thus must draw individuals to man them from other billets or from the Individual Ready Reserve (IRR). Examples of such headquarters include Multi-National Force–Iraq and Combined Security Transition Command–Afghanistan. This analysis focused on ordinary staff officer positions, not on positions with supervisory responsibilities such as directorate or division chiefs.

General Context

Three scenarios were developed at the individual officer level to represent alternative positions to be filled that require different skills and abilities. Respondents were told that they were the chiefs of staff of a 500-person multinational headquarters providing strategic direction to coalition military efforts in a small-scale counterinsurgency campaign. The three scenarios required the chief of staff to assign officers to emerging vacancies in the following directorates: operations, programs and resources, and strategic plans. While U.S. Army Human Resources Command assigned officers against these vacancies, the survey scenario informed respondents that they preferred to make the final decision themselves based on the officers' personal characteristics. Their staff identified candidates for each position from among the incoming officers. They interviewed each of the officers upon their arrival, and they all appeared competent.

Respondents

Either 105 or 106 qualified individuals answered each of the individual survey questions in each of the three scenarios.[1] Respondents to this survey had generally served in some sort of staff role. Table 5.1 describes the distribution by most responsible position held and average number of months deployed. In all, 106 individuals took this survey, with an average of 22.6 months deployed.

Individual Characteristics

Characteristics were derived from interviews with officers who had held responsible positions on various high-level staffs. Individual candidates differ along the characteristics described in the choice scenarios, and included:

Table 5.1
Individual Survey Sample Distribution by Most Responsible Position

Position	Active		Guard		Reserve	
	Count	Average Months Deployed	Count	Average Months Deployed	Count	Average Months Deployed
Other brigade-level command	1	48.0	2	12.5	2	11.5
G/C/J-3					1	30.0
Division or higher chief of staff	4	44.3	3	23.0	1	12.0
Deputy BCT commander			2	20.5		
Other battalion-level command			2	18.0		
Other division staff officer	7	43.4	11	18.4	9	32.6
Battalion/brigade staff officer	4	25.5	12	16.6	7	13.0
Transition team commander	1	40.0	9	21.3	5	19.4
Other	2	58.0	11	15.4	7	13.9
Total	19	41.4	52	17.9	32	20.1

---

[1] There were 105 responses for scenario 1, choices 1–3, and 106 for the rest of the scenarios or choices. Different respondents chose not to answer certain questions. The reason(s) for the missing responses is not clear. As such, all available information (i.e., the four answers provided by these respondents) was used in estimation.

- **Period of availability** (two levels): The amount of time for which the officer was available for assignment.
  - *Six months:* officer available for six months
  - *12 months:* officer available for 12 months.

- **Military branch** (two levels per scenario): The branch designation of the officer under consideration. The branch designations were spelled out by their title.
  - *Acquisition (51).* This branch represented a specialist background with regard to the programs and resources position and allowed the study team to assess the relative value of military education and training against civilian-acquired skills.
  - *Air defense artillery (14).* This branch represented a generalist background in the operations career field.
  - *Logistician (90).* This branch represented a generalist background in the logistics career field.

Only two military branch options were offered per scenario depending on the case.[2]

- **Prior deployment experience** (three levels): The last deployment experience the officer had doing related work:
  - two years ago
  - three years ago
  - four years ago.

- Force structure component (three levels):
  - USAR
  - ARNG
  - Regular Army.

- **Civilian-acquired skill** (three levels per scenario, ARNG and USAR only): Indicated whether an ARNG or USAR officer had an equivalent or relevant skill set, acquired through civilian training or employment. Some USAR and ARNG officers had no relevant civilian-acquired skills and were labeled as "none." Respondents were asked to assume that Regular Army officers have no relevant civilian-acquired skills and will be labeled as "N/A."
- *None (all scenarios):* Had no relevant civilian-acquired skills.

---

[2] In the description of the unit attributes shown to respondents, it was stated that military branch would also include military intelligence (35) and strategist (FA59). However, only acquisition, air defense artillery, and logistician were used in the actual experiment.

- **District manager for a small retail chain** (scenario 1): Managed several stores. Responsible for:
  - sales, costs, and profitability of stores
  - marketing and advertising within his district
  - implementing corporate initiatives.

- *Vice president of operations for a trucking company (scenario 1):* Directed the operations and activities of the trucking company. Responsible for:
  - scheduling and coordination
  - monitoring status of all pickups and deliveries across several states
  - scheduling maintenance.

- *Contracting manager for a small manufacturing firm (scenario 2):* Managed purchasing operations, vendors' reliability, and cost. Responsible for:
  - developing and sustaining a network of suppliers for raw materials, components (e.g., fuses, assemblies)
  - developing and sustaining a network for support services (e.g., waste disposal, accounting).

- **District contracting official, U.S. Army Corps of Engineers** (scenario 2): Very experienced in managing contract operations. Responsible for:
  - soliciting contract support for construction, professional support, and administrative support for an engineer district
  - working within the framework provided by the Federal Acquisition Regulation (FAR).

- *Assistant district attorney (scenario 3):* Supervised the development of criminal cases. Responsible for:
  - coordinating jurisdiction and investigations with federal, state, and local law enforcement agencies
  - helping set priorities for enforcement and prosecution
  - remaining keenly aware of public perceptions of law enforcement efforts' fairness and legitimacy.

- *Police lieutenant in a medium-sized city (scenario 3):* Supervised and coordinated the investigation of criminal cases with other police divisions and jurisdictions. Responsible for:
  - maintaining logs, preparing reports, and directing the preparation, handling, and maintenance of departmental records.

Only three civilian-acquired skill options, including "none," were offered per scenario depending on the case. Regular Army officers under consideration were assumed not to have a civilian-acquired skill, and thus this characteristic was labeled "not applicable" or "N/A." Appendix A provides the survey instruments for the individual officer level, including the full description of the scenarios and unit characteristics.

Results from Individual Scenario 1: Desk Officer for Operations Directorate

The task of scenario 1 required respondents to choose an officer to replace a regional command desk officer, an O-4 position. The Director of Operations (CJ-3) created these desk officer positions to monitor developments in the command's four regional commands.

Respondents were told that the description of responsibilities for this position included:

- reporting to you and your commander on ongoing activity at any given moment in the regional command for which the desk officers are responsible
- providing updates on whether the coalition and host-nation government are winning or losing, and why
- monitoring military operations undertaken by your subordinate combined joint task force (CJTF) and any significant air operations, as well as independent actions taken by other U.S. government agencies, the host-nation government, and nongovernmental organizations.

The required skills for the position were listed as:

- must be able to navigate U.S. military processes
- should understand how theater operations might affect popular support for the coalition and host-nation government.

The first data column of Table 5.2 (Model 1) presents the results from the choice experiment for this scenario using the complete sample. All characteristics entered the model as indicator (or "dummy") variables that take on a value of one if a characteristic takes the value indicated in the "Variables" column, and zero otherwise. The reference case (for which all indicator variables are assigned a value of zero)[3] was a Regular Army officer (thus with no applicable civilian-acquired skill) available for six months, in the logistics branch, with prior deployment experience four years ago.

The results show that, all else equal, respondents tended to choose Regular Army officers (note the negative and significant coefficients on USAR and ARNG variables). There were no discernible preferences between the ARNG and USAR, however. In addition, in this experiment, there were no statistical differences in respondents' preferences based on the branch of the candidate. Respondents preferred candidates with

---

[3] This corresponds to a scale index value of zero.

Table 5.2
Choice Experiment Results, Individual Scenario 1: Desk Officer for Operations Directorate

Variables	Model 1	Model 2
Skill—vice president of trucking (USAR/ARNG only)	2.009***	2.008***
	(0.219)	(0.220)
Skill—district manager retail (USAR/ARNG only)	1.972***	1.984***
	(0.250)	(0.255)
USAR	−0.852***	−0.391
	(0.234)	(0.245)
ARNG	−0.812***	−0.352
	(0.278)	(0.292)
Availability—12 months	0.859***	0.847***
	(0.163)	(0.168)
Branch: Air defense artillery	0.0787	0.0946
	(0.100)	(0.106)
Prior Deployment Experience—three years ago	0.334**	0.337**
	(0.132)	(0.138)
Prior deployment experience—two years ago	0.212	0.203
	(0.236)	(0.241)
Regular Army respondent*USAR	—	−1.681***
		(0.407)
Regular Army respondent*ARNG	—	−1.609***
		(0.396)
Number of observations	106	106
Number of choices	1,581	1,581

NOTES: Multinomial logit model results for scenario 1 with clustered (by individual) standard errors in parentheses. *** $p < 0.01$, ** $p < 0.05$, * $p < 0.1$. Baseline category is Regular Army officer (with no applicable civilian-acquired skill) available for six months, in the logistics branch, with prior deployment experience five years ago.

longer periods of availability, which could outweigh preferences for Regular Army officers.

As in the previous contexts, we tested for systemic differences in response between Regular Army, ARNG, and USAR respondents. As seen on the interaction variables for Model 2, Regular Army respondents were less likely to select ARNG or USAR candidates (equally across USAR and ARNG), but ARNG and USAR respondents did not tend to discriminate on this attribute.[4]

Results from Individual Scenario 2: Programs and Resources Directorate

The task of scenario 2 required respondents to choose an officer replacement in the programs and resources directorate, an O-5 position, at the same theater-level command. The command's objective was to develop a robust host nation and regional base of vendors to provide bulk supplies, as well as transportation and maintenance services to host-nation forces after U.S. forces transfer primary responsibility for security operations.

Respondents were told that the description of responsibilities for this position included:

- planning and coordinating centralized contract logistics support for U.S. and host-nation forces
- mentoring and training host-nation officers for subsequent transfer of this effort to the host nation.

The required skills for the position were listed as:

- ability to assess local vendors' existing capacity, reliability, and trustworthiness and their potential to improve
- understanding of existing host-nation laws and policies and the ability to think beyond the usual U.S. supply chain.

The first data column of Table 5.3 presents the results from the choice experiment for this scenario using the complete sample. The reference case (for which all indicator variables are assigned a value of zero) was a Regular Army officer (thus with no applicable civilian-acquired skill) available for six months, in the logistics branch, with deployment experience five years ago.

As in the previous individual officer scenario, respondents tended to value relevant civilian-acquired experience and 12 months of availability. Respondents did not discriminate based on candidates' branches. Unlike the first individual scenario, however, there is no evidence that even Regular Army respondents preferred Regular Army officers for this position, conditional on the skills possessed by the potential replacement officer. Rather, differences were manifest in strength of preference regarding the skills,

---

[4]   A Wald test of all excluded interaction terms in Model 2 has a Chi-squared test statistic of 4.98 (6 degrees of freedom) with p value of 0.55.

Table 5.3
Choice Experiment Results, Individual Scenario 2: Programs and Resources Directorate

Variables	Model 1	Model 2
Skill—contracting manager, small firm (USAR/ARNG only)	2.758***	3.100***
	(0.295)	(0.331)
Skill—district contracting manager, U.S. Army Corps of Engineers (USACE) (USAR/ARNG only)	2.649***	2.863***
	(0.306)	(0.349)
USAR	−0.209	−0.212
	(0.221)	(0.223)
ARNG	0.114	0.119
	(0.159)	(0.162)
Availability—12 months	0.448*	0.457**
	(0.231)	(0.232)
Branch: Acquisition	−0.0658	−0.0891
	(0.135)	(0.134)
Prior deployment experience—three years ago	0.442*	0.429*
	(0.252)	(0.253)
Prior deployment experience—two years ago	−0.0307	−0.0615
	(0.325)	(0.326)
Regular Army respondent*contracting manager, small firm		−1.271***
		(0.464)
Regular Army respondent*district contracting manager, USACE		−0.759*
		(0.412)
Number of observations	106	106
Number of choices	1,590	1,590

NOTES: Multinomial logit model results for scenario 2 with clustered (by individual) standard errors in parentheses. *** $p < 0.01$, ** $p < 0.05$, * $p < 0.1$. Baseline category is Regular Army officer (with no applicable civilian-acquired skill) available for six months, in the logistics branch, with prior deployment experience four years ago.

with Regular Army respondents placing somewhat less weight on civilian-acquired skills.[5] Model 2 presents these results.

---

[5] *P*-value of the Wald test associated with excluded interactions is 0.288 with Chi-squared test statistic of 7.37 (6 degrees of freedom).

Results from Individual Scenario 3: Counternarcotics Planner, Directorate of Strategic Plans and Policy

The task of scenario 3 required respondents to choose an officer to replace the counternarcotics planner, directorate of strategic plans and policy (CJ-5) vacancy, an O-5 position. Respondents were told that counternarcotics efforts were nominally led by officials from the U.S. State Department, but the lucrative narcotics trade and the associated criminal networks have significant impacts on security efforts. Insurgents derived much of their revenue from trafficking in narcotics, and use criminal networks to move money, people, and weapons.

The description of responsibilities for this position included:

- planning and coordinating intelligence, surveillance, reconnaissance, and security activities in support of counternarcotics operations
- developing host-nation security forces, including general and special police forces responsible for aspects of the counternarcotics mission.

The required skills for the position were listed as:

- developing the command's counternarcotics strategy
- coordinating with the local embassy, host-nation government, the United Nations, and representatives of allied governments counternarcotic operations.

Initial estimation of the standard choice model for scenario 3 resulted in coefficient estimates with extremely large standard errors.[6] After further investigation, it was discovered that 97.5 percent of chosen alternatives involved selection of an individual with a civilian skill (either police lieutenant or assistant district attorney).[7] Because all Regular Army individuals were assumed to not have similar civilian-acquired skills, this also means that 97.5 percent of chosen individuals for this scenario were in the ARNG or USAR. Statistically, this implies that a relevant civilian-acquired skill is a dominant attribute; that is, this variable explains a very large proportion of respondent choices. This result also precluded precise estimation of the coefficients on the other attributes besides component.[8] Nevertheless, from this simple result, we may conclude that, on average, respondents preferred either ARNG or USAR with civilian skills to Regular Army component individuals, regardless of the respondent's component.

---

[6]   This model assumed nonclustered standard errors.

[7]   Regular Army respondents made this choice 95 out of 100 times (95 percent), while ARNG and USAR respondents made this choice 422 out of 430 times (98 percent).

[8]   The reason is that statistical identification of the other coefficients relies on variation in choices within a choice occasion.

Qualitative Analysis

After completing the choice experiment, respondents were asked to describe (in a couple of sentences) "who was the most suited for this vacancy." The goal of these questions was to understand which additional characteristics not included into the choice experiment mattered when selecting among the candidates. Similar to the unit survey, the comments fell into three types of categories:

- those reiterating the attributes already listed in the experiment
- comments that identified new skills or individual characteristics
- comments asking for more information about the scenario or attributes.

Table 5.4 shows the distribution of comments among these categories by scenario. Comments included into the "New Attribute" category comprised between 23 to 50 percent of the sample. For scenario 1, these comments focused on candidates' analytical skills and other intangible characteristics that are usually assessed during face-to-face interviews rather than conveyed in resumes. These characteristics fell into the following categories:

- analytical skills and the ability to handle complexity (e.g., "The individual must be able to deal with a multitude of streams of data while also understanding military operations. They also must be able to understand the implications of military activities on the civilian sector")
- the ability to see a big picture (e.g., "holistically assess the operations effect on the civilian population [determine win/lose] based on the experience of for profit employment")
- networking skills (e.g. "Previous experience in a high-stress environment—can speak truth to power. Possesses ability to network to get things done—not arrogant about his rank, position, ratings, education")
- communication skills ("Must be a good communicator")

Table 5.4
Distribution of Comments Across Categories for Individual Survey

	Response Rate	Redundant	New Attribute	Scenario
Scenario 1	90% (68)	50% (34)	50% (34)	0%
Scenario 2	86% (65)	68% (44)	31.5% (20)	1.5% (1)
Scenario 3	88% (66)	75.5% (50)	23% (15)	1.5% (1)

NOTES: Raw numbers of responses in parentheses. Response rate = comments reiterating the importance of attributes already listed in the choice experiment; new attribute = comments mentioning attributes not listed in the choice experiment; scenario = comments on the clarity or plausibility of the scenario.

- substantive knowledge of the region and army operations (e.g., "Must understand Army operations")
- multitasking capabilities (e.g., "Someone open minded, can set aside preconceived perceptions, that can multitask and from the information provided reach sound logical conclusions").

For scenario 2, there were fewer comments regarding new attributes; most respondents felt that there was a good match between the hypothetical candidates included in the choice experiment and job description. The comments focused on candidates' knowledge of the country and the ability to sense corruption, familiarity with FAR, and commitment to getting a job done (e.g., "Highly experienced, even tempered with a broad perspective of private sector logistics, transportation. Skilled in recognizing corruption, theft and mismanagement"; "Individual must possess the ability to rapidly assess vendor capabilities given the current and projected operational environment").

For scenario 3, comments regarding new attributes focused on the following skills:

- interpersonal skills and coordination (e.g., "Someone who can work with a variety of agencies and keep the cats all going in the same direction. A person who can interact and motivate individuals with varying interests and agendas to keep everyone moving toward the overall agreed to goal"; "Must have ability to develop and maintain relationships with local LE [law enforcement]")
- education (e.g., "For any planner billet especially O-4 and above you have to seek a SAMS [School of Advanced Military Studies] or SAW [School of Advanced Warfighting] graduate." Experience in money exchanges [legal or illicit]; will accept previous studies on international banking and hawalas—prefer a liberal arts degree—good with math)
- knowledge of the host nation and language skills (e.g., "can speak the targeted foreign language)
- negotiation skills (e.g., "Someone who has experience dealing with a diverse set of individuals and who is used to cobbling together agreements, compromises and collaborative efforts is best").

The diversity of comments regarding additional attributes suggests that there was no single characteristic that influenced respondents' choices systematically but was not included in the choice set.

Findings

Regular Army Respondents Preferred Regular Army Candidates for Operations Directorate Positions

Regular Army respondents preferred Regular Army officers for this position, *when ARNG or USAR options lacked significant levels of relevant civilian-acquired skills.* In our interviews, respondents had indicated that officers working in the operations directorate needed to have a high level of understanding of military operations and staff processes. When ARNG and USAR candidates had high levels of relevant civilian-acquired skills, as indicated by their civilian status as a district manager or a trucking company vice president, even Regular Army respondents tended to select them. Ergo, analytic results suggest that proficiency in understanding military operations and staff processes is paramount, but that Regular Army officers believe that high levels of general skills in monitoring and managing day-to-day operations can compensate for their absence.

Respondents Generally Valued Civilian-Acquired Skills over Other Factors in Selecting Individual Staff Officers

In these scenarios, ARNG and USAR soldiers' civilian-acquired skills significantly outweighed all other factors in explaining respondents' choices. For example, Regular Army respondents preferred ARNG or USAR candidates who were district managers or trucking company vice presidents to Regular Army candidates with no relevant civilian-acquired skills. The study team developed these options because we felt that both these positions would have some experience monitoring and tracking operations, albeit in a considerably different context from counterinsurgency operations. Regular Army respondents generally preferred candidates with these attributes to all others. Civilian-acquired skills were more clearly relevant to the position described in both the plans and resources directorate and the counternarcotics planner position.

Our analysis therefore indicates that respondents placed a high value on civilian accomplishment. If the skills are transferable, as in the scenario in which respondents could choose an ARNG or USAR Soldier for the plans and resources directorate who was a contracting specialist for the Corps of Engineers in civilian life, the preference was even more pronounced. These models were consistent with the qualitative data obtained in the course of developing the survey instruments. Respondents often highlighted examples in which ARNG and USAR officers with such skills turned out to be more valuable in many roles within the complex political-military context of a strategic headquarters than some of their more conventionally trained Regular Army counterparts.

This analysis should not be taken to imply that ARNG or USAR officers in general are preferred to Regular Army officers because of their civilian-acquired skills. As ARNG and USAR officials are keenly aware themselves, reservists may or may not have an impressive job title or relevant civilian-acquired skill. When they do, however, it seems clear that even Regular Army respondents valued those skills in support of the military mission.

Continuity Is an Important Attribute

Aside from civilian-acquired skills, the most important attribute for the programs and resources (C-8) position was continuity. Respondents preferred candidates who would be available for 12 months to those who would be available for shorter spans of time. Continuity was also important in selecting candidates for the C-3 position. All Army respondents accorded similar value to this attribute.

Attributes Other Than Component Can Be More Important

As with the analysis of the unit results, our analysis of the individual survey data suggests that while Regular Army respondents may generally prefer Regular Army candidates for staff positions—although not always—they value ARNG and USAR officers under many conditions. Obviously, those conditions include civilian-acquired skills, but they also include longevity in the position and sometimes prior deployment experience. This study has probably omitted several relevant attributes. The larger message, however, is that individual attributes other than component are valued by all Army respondents.

Conclusion

This chapter explained our assessment of the degree to which Army respondents valued various individual attributes, including component, for high-level staff billets. Regular Army respondents preferred Regular Army candidates for positions in which primary responsibilities included monitoring tactical operations and navigating standard military processes. For other staff positions, this preference was considerably less pronounced and even nonexistent in some cases. Respondents preferred candidates with relevant civilian-acquired skills to other candidates, including Regular Army candidates, all else being equal. The preference for relevant civilian-acquired skills should not be taken as a preference for ARNG or USAR officers generally, since there is no guarantee that available candidates will possess the desired skills.

In the next chapter, we will explore the potential implications of the findings in the previous chapters for Army force planning.

CHAPTER SIX

Implications

The purpose of this study was to identify and assess the perceived relative importance of potential determinants of operational effectiveness in counterinsurgency operations for Army units and individuals among Army leaders. The central policy-relevant question, however, concerned the degree to which the Army could rely on ARNG and USAR units to fill certain roles, under certain conditions in a counterinsurgency environment, such as those of Afghanistan and Iraq from 2001 to 2015. To the extent that this analysis is valid, it is valid only for similar contexts. The research design did not address other operational themes, notably traditional combat operations.

The previous chapters reported the results of a series of choice experiments designed to uncover preferences of Army officers about the characteristics of units and individuals. Missions, threat levels, the operational environment, and job descriptions (in the individual case) were varied across scenarios to represent a range of potential experiences and test consistency of implied preferences. Qualified individuals (Regular Army, ARNG, and USAR colonels) were asked multiple questions per scenario that were experimentally designed to facilitate statistical estimation of models that predicted the observed choices. The questions asked respondents to choose a preferred unit or individual to perform a particular mission from among a set, and the results uncovered the perceived relative values of the included characteristics. This chapter explores the implications of those findings for Army force planning.

Review of Key Findings

Following a decade of shared operational experience, Regular Army respondents generally valued Regular Army units over their ARNG and USAR counterparts in a counterinsurgency environment, especially for high-threat scenarios involving close combat. In general, this preference was not shared by ARNG and USAR officers. In either case, we assume their choices reflected their professional assessments of the relative capabilities of these units in the contexts described, although other factors could also be in play. This is the most solidly grounded finding of the study, and indicates that well over a decade of shared combat experience could not overcome traditional differences in perspective between leaders in the Regular Army and Reserve components. Bridging this gap will continue to warrant Army leaders' attention, as it has in the past.

These results do not necessarily carry over to enabler units and individual positions. Regular Army respondents did not always choose Regular Army units. Indeed, in many cases, combinations of other factors and attributes outweighed component preferences. Although the study cannot explain the reason for these differences in preferences, one possible explanation is that the Regular Army, ARNG, and USAR comprise different combinations of forces and capabilities, and their units were typically employed in different roles. Regular Army maneuver units were typically employed against the higher threats, so their officers developed a perspective of war that likely differs from those of other Army personnel who did not have similar experiences. As a result, the "component" attribute of the exercise is used as a proxy for unit effectiveness.

From the standpoint of Regular Army officers, risk was far more likely to be acceptable under conditions in which all respondents consider risk levels to be acceptable. In short, it would seem more prudent to employ ARNG or USAR capabilities under conditions that all Army respondents would agree to entail acceptable risks than to employ them in contexts in which only ARNG or USAR respondents view the risks as acceptable.

Of course, as we will explain further below, that Regular Army respondents' tendency to choose Regular Army units pertains mostly to maneuver battalions and BCTs, and individuals working in domains dominated by military processes, for example, in the operations directorate. Our analysis of the survey data indicated little tendency to select Regular Army over other component enabler units. Regular Army respondents tended to be only slightly more likely to choose Regular Army candidates for individual staff positions. In the case of staff positions, significant levels of relevant civilian-acquired skills could cause Regular Army respondents to value ARNG or USAR candidates over Regular Army alternatives.

Unit Surveys

Regular Army respondents chose to employ Regular Army maneuver units in most scenarios described in this analysis, all else equal. ARNG and USAR officers did not share that view. That Regular Army valuation was even more pronounced with respect to high-risk scenarios. It is consistent with an assessment that Regular Army maneuver units operate at a higher level of effectiveness than their less well trained and less experienced ARNG counterparts. ARNG respondents did not demonstrate that same preference, however.

Regular Army respondents did not demonstrate a similar tendency with respect to enabler units such as engineer battalions and MP companies. Their tendency toward indifference to component may be a function of the low number of respondents to the engineer survey or the nature of the engineer mission in counterinsurgency. The few

Regular Army engineers who took the survey did seem to choose Regular Army engineer companies somewhat more frequently, however.

Regardless of respondents' component, other factors—alone or in combination—could outweigh units' component status for respondents. Operational risk or continuity (both measures of opportunity cost prior experience in theater, unit type)—for example, armored units in scenarios in which hard fighting was expected and high levels of predeployment training—could mitigate ARNG status, even for Regular Army officers. Ergo, it seems that it is possible to mitigate maneuver units' perceived deficiencies through some combination of additional predeployment training and in-country experience. Notably, the six months of predeployment training that represented the high end of options in our choice experiment was insufficient to outweigh Regular Army respondents' assessment that Regular Army maneuver units were more effective, with all other attributes being equal.

Of the other factors, operational risk and continuity were the most important. We expressed opportunity costs in terms of operational risk for the BCT scenarios and in terms of continuity in the area of operations for maneuver battalions.[1] These expressions of options' inherent opportunity costs were the most important variables for all respondents in all scenarios. Keeping risk low—or continuity high—was more important to Regular Army respondents than employing a Regular Army BCT in low- and moderate-threat scenarios. In these scenarios, Regular Army respondents decided that employing ARNG maneuver units incurred fewer risks than diverting Regular Army maneuver units from other higher-priority tasks. In high-risk scenarios in which heavy combat was anticipated, Regular Army respondents generally valued Regular Army maneuver units over their ARNG counterparts.

Leadership—or, more accurately, key unit leaders' proficiency with regard to the C2 of military operations—emerged as a very important, but largely unexplored, factor in determining units' operational effectiveness. In the interviews and focus groups the research team conducted to support development of survey instruments, respondents frequently explained the difference in unit capabilities in terms of key unit leaders' experience and expertise in the conduct of military operations. We did not include this characteristic in most surveys, however, because of its highly subjective nature. Instead, respondents were asked to assume that all other characteristics between unit choices were essentially unknown. It is possible, although not provable with the information we have, that expectations about leadership characteristics are captured in the estimated differences between Army respondent choices.

---

[1]   We expressed cost in a number of different ways in the different surveys, because we could not be sure which measure would resonate with respondents.

Individual Survey

The survey of individuals assessed the degree to which the Army could rely on ARNG and USAR officers to fill key billets in contingency headquarters. Such headquarters as Multi-National Force–Iraq and the Combined Security Transition Command–Afghanistan, among others, typically have broader responsibilities than the C2 of military operations. Typical responsibilities include building partner security forces' capabilities and capacity and supporting political and economic development. Given the broader and somewhat unfamiliar focus, newly assigned staff officers typically have to adapt and adjust to their responsibilities regardless of their component. We assessed the relative importance of various characteristics, including component, in three different contexts: monitoring operations and managing contract support and strategic planning in support of counternarcotics activities.

Regular Army respondents generally chose Regular Army candidates for positions in the operations directorate, in which familiarity with military operations and processes was particularly important. Reserve component respondents did not follow a similar pattern. Furthermore, there was no corresponding preference across the other positions, which notionally required officers to familiarize themselves with nonmilitary issues, capabilities, and processes.

Respondents typically accorded the highest weight to candidates' relevant civilian-acquired skills. For example, all Army respondents typically preferred ARNG and USAR candidates whose notional civilian job was either as a trucking company vice president or a district manager—positions requiring the monitoring and control of business operations—to fill vacancies as desk officers in an operations directorate. They preferred ARNG and USAR officers with contracting and procurement experience for the position responsible for developing a partner nation's supplier base, and ARNG and USAR officers who were either prosecutors or police officials in civilian life to fill the counternarcotics planner position. As the foregoing list indicates, these civilian-acquired skills were relevant to the position in question. The operations candidates did not simply work in transport or retail; they had reached positions of significant responsibility. In actuality, while many ARNG and USAR officers possess significant levels of relevant civilian-acquired skills, many do not, and civilian-acquire skills in one area may not be relevant to other areas. Bluntly stated, it should not be assumed that all ARNG or USAR officers possess high levels of civilian-acquired skills that are relevant to the military position that they may fill.

As with the analysis of the unit results, our analysis of the individual survey data suggests that while Regular Army respondents may generally value Regular Army candidates for staff positions—although not always—they value ARNG and USAR officers under many conditions. Obviously, those conditions include having candidates with relevant civilian-acquired skills, but they also include potential longevity in the position and sometimes prior deployment experience. Although results may change with the inclusion of potentially omitted relevant attributes, the larger message is that component is often outweighed by other attributes for individual positions in higher-level staffs.

Method

A final conclusion regards the methods used in this report. Choice experiments (and more generally, the entire family of stated preference methods) were developed in the transportation and economics literatures as a means of valuing various attributes of private and public goods. To our knowledge, this is the first application of the choice experiment methodology in the context of expected military unit and individual performance, in which the underlying value of characteristics pertaining to hypothetical units or individuals is estimated. While this value is not placed in monetary terms, the experiments performed in this report used operational risk and continuity as proxies for the opportunity costs of a particular choice. Respondents had to determine how much risk to the overall campaign plan they were willing to accept to compensate for any (potential) perceived effectiveness differences. This is extremely important in inducing the types of trade-offs necessary for proper application of a stated preference exercise.

Despite the potential limitations in the method—chief among them the hypothetical nature of the experiments and the limitations on information that can be provided to a respondent—we believe that stated preference methods, including choice experiments, can help Army and other service leaders address key policy questions about the trade-offs inherent in many decisions. Potential applications include, but are certainly not limited to, estimating the sufficient or optimal level of training to certain units needed to fulfill ex ante expectations of performance; decisions regarding the allocation of units or individuals to given missions, threat environments, and operational environments; and estimating the relative value of services and subservices provided by a unit or individual. Examples of the latter might include situations in which a unit is involved in a multidimensional mission with several potential measures of success but with constrained resources that force commanders to make strategic trade-offs in resource allocations toward each of the goals. In short, when the policy question under consideration involves trade-offs between competing attributes coupled with limited resources yet little real-world data to analyze (because of, say, a novel context), choice experiments provide one potential method that can help shed light on the preferences and implied values of decisionmakers.

Caveats

Some caution must be taken with regard to these findings. On the one hand, the foregoing findings were the result of a valid statistical analysis of the available data. Our interviews may not have identified the most important determinants of operational effectiveness, however, or the appropriate range of variation. Most importantly, our sample sizes were very small, and subject to selectivity bias. Only about 200 of the thousands of Regular Army, ARNG, and USAR colonels and general officers elected to respond to the survey.[2] However, to the extent that respondents were representative of Army colonels and general officers, choice patterns accurately reflected their underlying preferences, and the estimated models accurately represented respondents' views about the relative importance of various factors in determining operational effectiveness at the unit and soldier level. We already acknowledged that Regular Army and ARNG/USAR respondents differ significantly with respect to the value of Regular Army maneuver forces in a counterinsurgency environment, regardless of the threat level.

But while there are grounds for caution, there are also grounds for qualified confidence in the results. The factors described in these paragraphs were found to be statistically significant, and both focus groups and postexperiment comments suggest that, on average, respondents took the exercise seriously, understood it, and cognitively engaged in trade-offs across attributes. Furthermore, the results of the quantitative analysis appear to be consistent with the information gleaned from our interviews and the focus groups we used to validate the survey instrument. So although sample sizes are small, they were sufficient to produce statistically valid models that more or less conform to widely shared perceptions emerging from our qualitative data.

Recommendations

The following recommendations derive from our findings. In overview, our recommendations generally follow a pattern of explicitly assessing the risk associated with different operational requirements, preserving sufficient Regular Army capacity to meet high-risk requirements, and taking measures to mitigate any residual risks associated with employing ARNG or USAR forces in lower-risk contexts.

Like our findings, the focus on measures mitigates the risks identified by Regular Army officers in the course of this study or implied by their responses to these surveys. These recommendations focus on improving the Army's operational effectiveness. As with the findings, recommendations apply primarily in the counterinsurgency context.

Consider Increasing Predeployment Training Period for ARNG Maneuver Units Conducting Counterinsurgency Operations

In one sense, all forces deployed to a counterinsurgency campaign conduct counterinsurgency operations. We use the term above as shorthand for units that will be responsible for synchronizing offense, defense, and stability operations within a defined area

---

[2] The low level of participation is somewhat understandable. The survey was necessarily time-consuming, complex, and somewhat artificial. Potential respondents are busy, consumed with important duties and with little discretionary time available in which to take a survey. Furthermore, as researchers, we were constrained to invite participation in the survey only once with limited opportunities to follow up.

of operations. Forces with such responsibilities have been known colloquially as "land-owners" over the past several years. We use the term *predeployment training* to connote the combination of pre- and postmobilization training reserve component forces undergo to prepare themselves for a specific deployment.

ARNG maneuver brigades with such responsibilities in 2004–2006 generally accumulated just a little less than six months of such training. Analysis of our survey results indicates that Regular Army respondents did not feel that six months of predeployment training outweighed the risk of employing ARNG maneuver forces, especially in high-threat scenarios.[3] This result indicates that even longer periods of peacetime and predeployment training might be necessary for maneuver units that will be responsible for an area of operations, although it cannot indicate how long such a period should be.[4]

Our analysis indicates that, if it is necessary to employ ARNG maneuver battalions and brigades to conduct counterinsurgency operations, some combination of additional annual training days, longer periods of mobilization, and more-robust predeployment training may be necessary. Alternatively, policymakers might decide to accept the risk associated with deploying less well-trained Army units to lethal environments. Finally, the Army might also consider deploying ARNG and USAR units in platoons and companies rather than battalions, brigades, and divisions to reduce requirements for predeployment training.

As with the foregoing recommendation, increasing requirements for annual and predeployment training would incur increased costs. The Army would need to maintain more ARNG and USAR force structure to generate and sustain the same level of rotational capacity. Generating that capacity would be more expensive on a per-unit basis, given the increased training day requirements.[5] More analysis would be required to determine whether it is more operationally and cost-effective to increase ARNG and USAR units' predeployment training or increase reliance on Regular Army forces.

Explore Options for Increasing Access to ARNG and USAR Soldiers with Relevant Civilian-Acquired Skills

Respondents preferred soldiers with civilian skills relevant to the positions for which they were being considered to those without, regardless of the soldier's component. This dynamic applied even in operations directorate positions primarily concerned with monitoring and managing military operations. The Army does not assign soldiers to military billets based on their civilian-acquired skills, or track soldiers' civilian skills,

---

[3] The experimental design did not allow for identification of the "break-even" level of training. This is a subject for future research.

[4] Klimas et al., 2014.

[5] Longer period of predeployment or postmobilization training implies longer periods of mobilization to achieve the same degree of operational capacity, leading to increased military personnel and training costs.

regardless of component. Previous DoD efforts to catalog military personnel's civilian-acquired skills have not proven fruitful to date, and have run into legal objections. The Army could increase its ability to identify potential operational requirements for soldiers with particular talents and incentivize soldiers with those skills to volunteer to fill positions in which those skills would be useful.

Assess the Utility of Increasing Key ARNG and USAR Leaders' Opportunities to Accrue More Operational Experience and Assign Regular Army Leaders to Command and Staff Billets in ARNG and USAR Units

Regular Army, USAR, and ARNG respondents agreed that key leaders' proficiency had a disproportionate effect on units' operational effectiveness, even though they might not have agreed about components' general levels of such proficiency. Proficiency in any field is mostly a function of practice in the relevant context. ARNG and USAR leaders generally accrue less relevant experience than their Regular Army counterparts. They spend less time deployed, less time training at home station, and experience fewer rotations at the Army's CTCs, the Army's premier substitutes for actual combat experience. Increasing the operational experience levels of key reserve component leaders may be one relatively low-cost way of increasing the effectiveness of the units they will lead. Doing so without disrupting existing systems would, of course, pose a significant challenge. Policymakers must, of course, consider costs in their analysis.

We offer the following ideas for the sake of illustration, not necessarily implementation. Determining the best approach to increasing reserve component leaders' relevant operational experience is worthy of careful study in its own right. The ideas include:

- **Continue to facilitate the transition of departing midgrade officers, warrant officers, and noncommissioned officers to the reserve components**. Such efforts are already underway as the Army draws down and indeed have always been a staple of transitions from active duty.[6]
- **Employ ARNG and USAR forces regularly on low-risk operational missions**. U.S. declaratory strategy posits a high level of military engagement globally, albeit focused more on strengthening relationships with partners and allies and building their capacity. To help meet this demand, the Army can mobilize reserve component units for up to 365 days to carry out preplanned missions in support of combatant commands.[7] In developing the Army's sourcing strategy to meet these demands, Army leaders should include the impact of sourcing decisions on leader development in the reserve components.

---

[6] Michelle Tan, "Army Launches New Incentives to Quit Active, Join Reserves: Guard and Reserve Test New Program for Drawing Active Duty Soldiers," *Army Times*, February 17, 2014.

[7] Lawrence Kapp and Barbara Salazar Torreon, *Reserve Component Personnel Issues: Questions and Answers*, Washington, D.C.: Congressional Research Service, June 13, 2014, p. 18.

- **Provide ARNG and USAR leaders with additional opportunities for active duty service.** Many ARNG and USAR soldiers serve on active duty but mostly for the purpose of supporting the administration, organization, and training of reserve component forces.[8] The Army could seek legislative authority to enable selected reserve component leaders to undertake broadening assignments with Regular Army forces. Implementing such a policy would prove challenging, as it would be necessary to find high-quality reserve component leaders who could afford to depart their civilian job for several years to serve with Regular Army units. Moreover, increasing opportunities for reserve component leaders could only come at the expense of decreasing opportunities for Regular Army leaders.
- **Assign Regular Army officers to key command and staff billets in ARNG and USAR units.** The foregoing recommendation to allow ARNG and USAR leaders to acquire experience in active component units is one method of increasing their leadership capabilities. Assigning Regular Army officers to key billets in reserve component units is another way of doing so. High-quality Regular Army officers could transmit the benefits of their operational experience to subordinates, peers, and even superiors in the ARNG and USAR. For this effort to have the intended benefit, however, such assignments would have to be high priority for both the active and reserve components. High-quality officers would have to consider such assignments an opportunity.

Explore Options for Increasing Utilization of Reserve Component Soldiers with Relevant Civilian-Acquired Skills

Respondents valued civilian-acquired skills greatly when evaluating individuals. For many positions, both Regular Army and reserve component respondents preferred reservists with relevant civilian acquired skills to Regular Army candidates. It is difficult to assess the degree to which the Army could rely on such soldiers could fill wartime requirements, however. Our survey merely indicates that such substitution may be possible in some cases; it does not indicate the full range of positions for which it may be possible or the number of reservists who might possess the desired skills. Currently, information on reservists' civilian employment is incomplete, unverified, and therefore unreliable.[9]

While the Army cannot compel ARNG and USAR soldiers with relevant civilian acquired skills to fill billets requiring those skills, it can incentivize them to volunteer for such assignments. As such, we recommend that the Army survey staff directors and chiefs of staff of various contingency headquarters and other nonstandard organiza-

---

[8]   Kapp and Torreon, 2014, p. 6.

[9]   Commission on the National Guard and Reserves, *Transforming the National Guard and Reserves into a 21st Century Operational Force: Final Report to Congress and the Secretary of Defense*, Washington, D.C.: U.S. Department of Defense, 2008, pp. 149–150.

tions identify the positions in which a reservist with relevant civilian-acquired skills might be preferred and the nature of the skills or experience in question. Concurrently, the Army should take steps to encourage soldiers to ensure their information in the DoD's Civilian Employment Information (CEI) database is complete and accurate. With information about potential demand and supply in hand, the Army can then determine the extent to which it can rely on soldiers with such skills to fill wartime billets. The Army should also investigate the kinds of incentives that would inspire ARNG and USAR soldiers to volunteer for such assignments.

Conclusion

This study investigated the expected operational performance of Army units and individuals, within the overall context of counterinsurgency operations. Based on limited data, current force structure, and the totality of our statistical results, the study team concludes that Regular Army officers assess that Regular Army maneuver forces are more appropriate for conducting combat operations in high-risk, high-threat areas of operation, in relation to ARNG forces. ARNG maneuver forces are assessed as appropriate for conditions of lower threat and risk, with appropriate preparation and other risk-mitigation measures. We should note those preparations would probably require longer periods of mobilization than the one year. Therefore, the Army should maintain enough Regular Army maneuver force capacity to meet these high-risk maneuver forces to the extent that counterinsurgency operations remain a planning priority. With adequate training and preparation, select ARNG and USAR enablers, such as engineers and military police, can be expected to perform well when deployed with appropriate predeployment training.

The results of the exercise also show that Regular Army respondents and their ARNG counterparts differ with respect to the expected effectiveness of Regular Army and ARNG maneuver units, while this difference is not as pronounced for enabling units and individuals. We are unable to explain this result, given the data in the experiment, although we note that this result is consistent with different experiences and therefore judgments of the respondents in the survey.

The degree to which these conclusions apply going forward is unclear. As noted, they apply primarily to the conduct of counterinsurgency operations. However, the 2012 Defense Strategic Guidance prescribed that "U.S. forces will no longer be sized to conduct large-scale, prolonged stability operations.[10] The extent to which these findings might be extrapolated into other types of operations is thus open to question. Certainly, traditional combat operations are generally assumed to pose greater risks and

---

[10] Leon Panetta, *Sustaining U.S. Global Leadership: Priorities for 21st-Century Defense*, Washington, D.C.: U.S. Department of Defense, January 2012.

require higher levels of proficiency from all arms. To the limited extent that it is possible to generalize from this study's focus on counterinsurgency to traditional combat, our analysis would indicate a need for increased reliance on Regular Army forces.

Survey Instruments and Conditional Logit Model

Appendix A, Survey Instruments, presents the survey instruments used to collect data for this report.

Appendix B, Conditional Logit Model, illustrates the general statistical model used to analyze the data.

These online appendixes are available for download at www.rand.org/t/RR1745.

Bibliography

Andrade, Dale, *Surging South of Baghdad: The 3d Infantry Division and Task Force Marne in Iraq, 2007–2008*, Washington, D.C.: U.S. Army Center of Military History, 2010.

Biddle, Stephen, *Military Power: Explaining Victory and Defeat in Modern Battle*, Princeton, N.J.: Princeton University Press, 2004.

Bransford, John, *How People Learn: Brain, Mind, Experience and Schools*, Washington, D.C.: National Academy Press, 1998.

Brown, John Sloan, "The Wehrmacht Mythos Revisited: A Challenge for Colonel Trevor N. Dupuy," *Military Affairs*, Vol. 51, No. 3, July 1987, pp. 146–147.

Carter, Phillip, and Nora Bensahel, "Reboot: Why the Army's Plan to Cut 80,000 Troops Doesn't Go Nearly Far Enough," *Foreign Policy*, June 26, 2013. As of March 7, 2015: http://foreignpolicy.com/2013/06/26/reboot/

Champ, Patricia A., Kevin J. Boyle, and Thomas C. Brown, eds., *A Primer on Nonmarket Valuation*, Dordrecht, the Netherlands: Kluwer Academic Publishers, 2003.

Commission on the National Guard and Reserves, *Transforming the National Guard and Reserves into a 21st Century Operational Force: Final Report to Congress and the Secretary of Defense*, Washington, D.C.: U.S. Department of Defense, 2008.

Connable, Ben, *Embracing the Fog of War: Assessment and Metrics in Counterinsurgency*, Santa Monica, Calif.: RAND Corporation, MG-1086-OSD, 2012. As of November 15, 2016: http://www.rand.org/pubs/monographs/MG1086.html

"Discrete Choice Modelling: Methods for Understanding Why People Make the Choices That They Do," Santa Monica, Calif.: RAND Corporation, RB-9204-RE, 2006. As of February 27, 2013: http://www.rand.org/pubs/research_briefs/RB9204.html

Dupuy, Trevor N., *Numbers, Predictions and War: Using History to Evaluate Combat Factors and Predict the Outcome of Battles*, New York: Bobbs-Merrill, 1979.

Ericsson, K. Anders, "The Influence of Experience and Deliberate Practice on the Development of Superior Expert Performance," in K. Anders Ericsson, Neil Charness, Paul J. Feltovich, and Robert R. Hoffman, eds., *The Cambridge Handbook of Expertise and Expert Performance*, Cambridge, UK: Cambridge University Press, 2006.

Ericsson, K. Anders, Ralf Th. Krampe, and Clemens Tesch-Römer, "The Role of Deliberate Practice in the Acquisition of Expert Performance," *Psychological Review*, Vol. 100, No. 3, 1993, pp. 363–406.

Feickert, Andrew, and Lawrence Kapp, *Army Regular Army (AC)/Reserve Component (RC) Force Mix: Considerations and Options for Congress*, Washington, D.C.: Congressional Research Service, R43808, December 5, 2014.

Freedberg, Sydney J., Jr., "Active vs. Guard: An Avoidable Pentagon War," *Breaking Defense*, June 28, 2013. As of March 7, 2015:
http://breakingdefense.com/2013/06/active-vs-guard-an-avoidable-pentagon-war/

Freedberg, Sydney J., Jr., "National Guard Commanders Rise in Revolt Against Active Army; MG Rossi Questions Guard Combat Role," *Breaking Defense*, March 11, 2014. As of March 7, 2015:
http://breakingdefense.com/2014/03/national-guard-commanders-rise-in-revolt-against-active-army-mg-ross-questions-guard-combat-role/

Gates, Robert M., "Utilization of the Total Force," memorandum for Secretaries of the Military Departments, Chairman of the Joint Chiefs of Staff, Under Secretaries of Defense, January 19, 2007.

Graham, David R., Robert B. Magruder, John R. Brinkerhoff, James L. Adams, Richard P. Diehl, Colin M. Doyle, and Anthony C. Hermes, *Managing Within Constraints: Balancing U.S. Army Forces to Address a Full Spectrum of Possible Operational Needs*, Alexandria, Va.: Institute for Defense Analyses, IDA Paper P-4579, September 2010.

Grass, Frank, "Authorities and Assumptions Related to Rotational Use of the National Guard," memorandum for Chief of Staff of the Army," May 31, 2013.

Grass, Frank, "Statement by General Frank J. Grass, Chief, National Guard Bureau, Before the Senate Armed Services Committee, 2nd Sess., 113th Cong., on Army Total Force Mix," Washington, D.C., April 8, 2014. As of March 7, 2015:
http://www.armed-services.senate.gov/imo/media/doc/Grass_04-08-14.pdf

Gronski, John L., Kurt Nielsen, and Alfred A. Smith, "2/28 BCT Goes to War," undated. As of April 7, 2015:
http://www.milvet.state.pa.us/PAO/pr/2006_07_01.htm

Headquarters, Department of the Army, "Counterinsurgency," Washington, D.C., FM 3-24, 2006.

Headquarters, Department of the Army, "The Brigade Combat Team," Washington, D.C., FM 3-90.6, 2006.

Headquarters, Department of the Army, "Military Police Operations," Washington, D.C., FM 3-39, 2010.

Headquarters, Department of the Army, "Engineer Operations," Washington, D.C., FM 3-34, August 2011.

Hensher, David A., John M. Rose, and William H. Greene, *Applied Choice Analysis: A Primer*, Cambridge, UK: Cambridge University Press, 2005.

Hoyos, David, "The State of the Art of Environmental Valuation with Discrete Choice Experiments," *Ecological Economics*, Vol. 69, No. 8, 2010, pp. 1595–1603.

Kanninen, Barbara J., "Optimal Design for Multinomial Choice Experiments," *Journal of Marketing Research*, Vol. 39, No. 2, February, 2002, pp. 214–227.

Kapp, Lawrence, and Barbara Salazar Torreon, *Reserve Component Personnel Issues: Questions and Answers*, Washington, D.C.: Congressional Research Service, June 13, 2014.

Klein, Gary, *Sources of Power: How People Make Decisions*, Cambridge, Mass.: Massachusetts Institute of Technology Press, 1999.

Klimas, Joshua, Richard E. Darilek, Caroline Baxter, James Dryden, Thomas F. Lippiatt, Laurie L. McDonald, J. Michael Polich, Jerry M. Sollinger, and Stephen Watts, *Assessing the Army's Active-Reserve Component Force Mix*, Santa Monica, Calif.: RAND Corporation, RR-417-1-A, 2014. As of November 15, 2016:
http://www.rand.org/pubs/research_reports/RR417-1.html

Lord, Robert G., and Karen J. Maher, "Cognitive Theory in Industrial and Organizational Psychology," in Marvin D. Dunnette and Leaetta M. Hough, eds., *Handbook of Industrial and Organizational Psychology*, Vol. 2, Palo Alto, Calif.: Consulting Psychologists Press, 1991, pp. 1–62.

Louviere, Jordan J., David A. Hensher, and Joffre D. Swait, *Stated Choice Methods and Applications*, New York: Cambridge University Press, 2000.

McFadden, Daniel, "Conditional Logit Analysis of Qualitative Choice Behavior," in Paul Zarembka, ed., *Frontiers in Econometrics*, New York: Academic Press, 1974, pp. 105–142.

Millett, Allan R., and Williamson Murray, *Military Effectiveness*, Vol. 1, *The First World War*, Winchester, Mass.: Allan and Unwin, 1988.

National Commission on the Future of the Army, *Report to the President and the Congress of the United States*, Washington, D.C., January 28, 2016.

Office of the Vice Chairman of the Joint Chiefs of Staff and Office of Assistant Secretary of Defense for Reserve Affairs, *Comprehensive Review of the Future Role of the Reserve Component*, Washington, D.C.: U.S. Department of Defense, 2011.

Panetta, Leon, *Sustaining U.S. Global Leadership: Priorities for 21st-Century Defense*, Washington, D.C.: U.S. Department of Defense, January 2012.

Peters, John E., Brian Shannon, and Matthew E. Boyer, *National Guard Special Forces: Enhancing the Contributions of Reserve Component Army Special Operations Forces*, Santa Monica, Calif.: RAND Corporation, TR-1199-A, 2012. As of November 15, 2016:
http://www.rand.org/pubs/technical_reports/TR1199.html

Pint, Ellen M., Matthew W. Lewis, Thomas F. Lippiatt, Philip Hall-Partyka, Jonathan P. Wong, and Tony Puharic, *Active Component Responsibility in Reserve Component Pre- and Postmobilization Training*, Santa Monica, Calif.: RAND Corporation, RR-738-A, 2015. As of July 17, 2017:
https://www.rand.org/pubs/research_reports/RR738.html

Reeves, Kevin, "155th BCT Hits the Ground Running," *Marine Corps News*, reprinted at GlobalSecurity.org, February 19, 2005. As of April 7, 2015:
http://www.globalsecurity.org/military/library/news/2005/02/mil-050221-usmc01.htm

Schroden, Jonathan, Rebecca Thomasson, Randy Foster, Mark Lukens, and Richard Bell, "A New Paradigm for Assessment in Counterinsurgency," *Military Operations Research*, Vol. 18, No. 3, 2013, pp. 5–20.

Tan, Michelle, "Army Launches New Incentives to Quit Active, Join Reserves: Guard and Reserve Test New Program for Drawing Active Duty Soldiers," *Army Times*, February 17, 2014. As of April 1, 2015:
http://archive.armytimes.com/article/20140217/NEWS/302170006/Army-launches-new-incentives-quit-active-join-reserves

United States Code, Title 32, Section 105, Inspection, August 23, 1992.

U.S. Department of Defense, "Managing the Reserve Components as an Operational Force," Washington, D.C.: Under Secretary of Defense for Personnel and Readiness, DoDD 1200.17, October 29, 2008.

U.S. Department of Defense, "Reserve Components Common Personnel Data System (RCCPDS): Reporting Procedures," Washington, D.C.: Under Secretary of Defense for Personnel and Readiness, DoD Manual (DoDM) 7730.54, October 17, 2013. As of September 1, 2017:
http://www.esd.whs.mil/Portals/54/Documents/DD/issuances/dodm/773054m_vol1.pdf

U.S. House of Representatives, *Howard P. "Buck" McKeon National Defense Authorization Act for Fiscal Year 2015: Report of the Committee on Armed Services House of Representatives on H.R. 4435 Together with Additional Views*, Washington, D.C.: Committee on Armed Services, H.R. 113-446, May 13, 2014, pp. 199–200.

U.S. Joint Chiefs of Staff, *Counterinsurgency*, Washington, D.C., Joint Publication 3-24, November 22, 2013.